Introduction to Financial Derivatives with Python

Introduction to Financial Derivatives with Python is an ideal textbook for an undergraduate course on derivatives, whether on a finance, economics, or financial mathematics program. As well as covering all of the essential topics one would expect to be covered, the book also includes the basis of the numerical techniques most used in the financial industry, and their implementation in Python.

Features
- Connected to a Github repository with the codes in the book. The repository can be accessed at https://bit.ly/3bllnuf
- Suitable for undergraduate students, as well as anyone who wants a gentle introduction to the principles of quantitative finance
- No pre-requisites required for programming or advanced mathematics beyond basic calculus.

Elisa Alòs holds a Ph.D. in Mathematics from the University of Barcelona. She is an Associate Professor in the Department of Economics and Business at Universitat Pompeu Fabra (UPF) and a Barcelona GSE Affiliated Professor. Her research focus has been on the applications of the Malliavin calculus and the fractional Brownian motion in mathematical finance and volatility modeling since the past fourteen years.

Raúl Merino has been working full-time in the industry as Risk Quant since 2008. He is also an Associate Professor at Pompeu Fabra University (UPF) where he teaches the course 'Financial Derivatives and Risk Management'. Raul holds a Ph.D. in Mathematics from the University of Barcelona. In his Ph.D., he studied the use of decomposition formulas in stochastic volatility models. His research interests are stochastic analysis and applied mathematics, with a special focus on applications to mathematical finance.

Chapman & Hall/CRC Financial Mathematics Series

Series Editors

M.A.H. Dempster
Centre for Financial Research
Department of Pure Mathematics and Statistics
University of Cambridge, UK

Dilip B. Madan
Robert H. Smith School of Business
University of Maryland, USA

Rama Cont
Department of Mathematics
Imperial College, UK

Robert A. Jarrow
Lynch Professor of Investment Management
Johnson Graduate School of Management
Cornell University, USA

Recently Published Titles

Stochastic Modelling of Big Data in Finance
Anatoliy Swishchuk

Introduction to Stochastic Finance with Market Examples, Second Edition
Nicolas Privault

Commodities: Fundamental Theory of Futures, Forwards, and Derivatives Pricing, Second Edition
Edited by M.A.H. Dempster, Ke Tang

Foundations of Qualitative Finance: Book 1: Measure Spaces and Measurable Functions
Robert R. Reitano

Introducing Financial Mathematics: Theory, Binomial Models, and Applications
Mladen Victor Wickerhauser

Foundations of Qualitative Finance: Book II: Probability Spaces and Random Variables
Robert R. Reitano

Financial Mathematics: From Discrete to Continuous Time
Kevin J. Hastings

Financial Mathematics: A Comprehensive Treatment in Discrete Time
Giuseppe Campolieti and Roman N. Makarov

Introduction to Financial Derivatives with Python
Elisa Alòs, Raúl Merino

For more information about this series please visit: https://www.crcpress.com/Chapman-and-HallCRC-Financial-Mathematics-Series/book series/CHFINANCMTH

Introduction to Financial Derivatives with Python

Elisa Alòs
Pompeu Fabra University, Spain

Raúl Merino
Pompeu Fabra University, Spain

Foreword by Frido Rolloos

CRC Press
Taylor & Francis Group
Boca Raton London New York

CRC Press is an imprint of the
Taylor & Francis Group, an **informa** business

A CHAPMAN & HALL BOOK

First edition published 2023

by CRC Press
6000 Broken Sound Parkway NW, Suite 300, Boca Raton, FL 33487-2742

and by CRC Press
4 Park Square, Milton Park, Abingdon, Oxon, OX14 4RN

© 2023 Elisa Alòs and Raúl Merino

CRC Press is an imprint of Taylor & Francis Group, LLC

Reasonable efforts have been made to publish reliable data and information, but the author and publisher cannot assume responsibility for the validity of all materials or the consequences of their use. The authors and publishers have attempted to trace the copyright holders of all material reproduced in this publication and apologize to copyright holders if permission to publish in this form has not been obtained. If any copyright material has not been acknowledged please write and let us know so we may rectify in any future reprint.

Except as permitted under U.S. Copyright Law, no part of this book may be reprinted, reproduced, transmitted, or utilized in any form by any electronic, mechanical, or other means, now known or hereafter invented, including photocopying, microfilming, and recording, or in any information storage or retrieval system, without written permission from the publishers.

For permission to photocopy or use material electronically from this work, access www.copyright.com or contact the Copyright Clearance Center, Inc. (CCC), 222 Rosewood Drive, Danvers, MA 01923, 978-750-8400. For works that are not available on CCC please contact mpkbookspermissions@tandf.co.uk

Trademark notice: Product or corporate names may be trademarks or registered trademarks and are used only for identification and explanation without intent to infringe.

Library of Congress Cataloging-in-Publication Data

Names: Alòs, Elisa, author. | Merino, Raúl, author.
Title: Introduction to financial derivatives with Python / Elisa Alòs, Pompeu Fabra University, Spain, Raúl Merino, Pompeu Fabra University, Spain.
Description: Boca Raton : Chapman & Hall, CRC Press, 2023. | Series: Chapman & Hall/CRC financial mathematics series | Includes bibliographical references and index.
Identifiers: LCCN 2022037925 (print) | LCCN 2022037926 (ebook) | ISBN 9781032211039 (hardback) | ISBN 9781032211053 (paperback) | ISBN 9781003266730 (ebook)
Subjects: LCSH: Derivative securities. | Python (Computer program language)
Classification: LCC HG6024.A3 A455 2023 (print) | LCC HG6024.A3 (ebook) | DDC 332.64/57--dc23/eng/20220923
LC record available at https://lccn.loc.gov/2022037925
LC ebook record available at https://lccn.loc.gov/2022037926

ISBN: 978-1-032-21103-9 (hbk)
ISBN: 978-1-032-21105-3 (pbk)
ISBN: 978-1-003-26673-0 (ebk)

DOI: 10.1201/9781003266730

Typeset in LM Roman
by KnowledgeWorks Global Ltd.

Publisher's note: This book has been prepared from camera-ready copy provided by the authors.

In loving memory of my father.

To my parents, Dani and Laura.

Contents

List of Figures — xiii

Foreword — xvii

Preface — xxi

CHAPTER 1 ▪ Introduction — 1

    1.1 FINANCIAL MARKETS — 1
    1.2 DERIVATIVES — 1
    1.3 TIME HAS A VALUE — 2
    1.4 NO-ARBITRAGE PRINCIPLE — 5
    1.5 CHAPTER'S DIGEST — 7
    1.6 EXERCISES — 7

CHAPTER 2 ▪ Futures and Forwards — 9

    2.1 FORWARD CONTRACTS: DEFINITIONS — 9
    2.2 FUTURES — 11
    2.3 WHY TO USE FORWARDS AND FUTURES? — 14
    2.4 THE FAIR DELIVERY PRICE: THE FORWARD PRICE — 15
        2.4.1 The General Approach — 15
        2.4.2 Some Special Cases — 17
            2.4.2.1 Assets that Provide a Known Income — 17
            2.4.2.2 Assets that Provide an Income Proportional to Its Price — 19
        2.4.3 The Price of a Forward Contract — 21

	2.4.4	The general case	21
		2.4.4.1 The Case of a Known Income	22
		2.4.4.2 Assets that Provide an Income Proportional to Its Price	22
2.5	CHAPTER'S DIGEST		23
2.6	EXERCISES		23

CHAPTER 3 ▪ Options — 25

3.1	CALL AND PUT OPTIONS	25
3.2	THE INTRINSIC VALUE OF AN OPTION	27
3.3	SOME PROPERTIES OF OPTION PRICES	27
	3.3.1 The Price of an Option vs the Price of an Asset	28
	3.3.2 The Role of the strike price	29
	3.3.3 The Role of the Price of the Underlying Asset	29
	3.3.4 The Role of Interest Rates	30
	3.3.5 The Role of Volatility	31
	3.3.6 The Role of Time to Maturity	31
	3.3.7 The Put-Call Parity	32
3.4	SPECULATION WITH OPTIONS	33
3.5	SOME CLASSICAL STRATEGIES	35
	3.5.0.1 Bull Spread	35
	3.5.0.2 Bear Spread	36
3.6	DRAW YOUR STRATEGY WITH PYTHON	37
3.7	CHAPTER'S DIGEST	42
3.8	EXERCISES	43

CHAPTER 4 ▪ Exotic Options — 45

4.1	BINARY OPTIONS	45
4.2	FORWARD START OPTIONS	46
	4.2.1 Compound Options	47
4.3	PATH-DEPENDENT OPTIONS	47

		4.3.1	Barrier Options	48
		4.3.2	Lookback Options	50
		4.3.3	Asian Options	51
	4.4	SPREAD AND BASKET OPTIONS		52
	4.5	BERMUDA OPTIONS		53
	4.6	CHAPTER'S DIGEST		53
	4.7	EXERCISES		53

CHAPTER 5 ▪ The Binomial Model — 55

	5.1	THE SINGLE-PERIOD BINOMIAL MODEL		55
		5.1.1	Relationship between European Options and Their Underlying in the Binomial Model	59
		5.1.2	Replication Portfolio for European Options	60
		5.1.3	The Risk-neutral Valuation	64
		5.1.4	Link the Model to the Market	67
	5.2	THE MULTI-PERIOD BINOMIAL MODEL		69
		5.2.1	Adjusting the Parameters	72
		5.2.2	Pricing a European Option	74
			5.2.2.1 Extended Framework	74
			5.2.2.2 Simplified Framework	84
		5.2.3	Early Exercise	84
	5.3	THE GREEKS IN THE BINOMIAL MODEL		87
		5.3.1	Delta	90
		5.3.2	Gamma	90
		5.3.3	Theta	91
		5.3.4	Vega	91
		5.3.5	Rho	92
		5.3.6	Approximating the Price Function	92
	5.4	CODING THE BINOMIAL MODEL		93
	5.5	CHAPTER'S DIGEST		105
	5.6	EXERCISES		106

Chapter 6 ▪ A Continuous-time Pricing Model   109

6.1	CREATING SOME INTUITION	109
6.2	THE BLACK-SCHOLES-MERTON FRAMEWORK	113
6.3	THE BLACK-SCHOLES-MERTON EQUATION	114
6.4	THE BLACK-SCHOLES-MERTON FORMULA	116
6.5	THE BLACK-SCHOLES-MERTON MODEL FROM A PROBABILISTIC PERSPECTIVE	120
6.6	THE BLACK-SCHOLES-MERTON PRICE AND THE BINOMIAL PRICE	126
6.7	THE GREEKS IN THE BLACK-SCHOLES-MERTON MODEL	127
	6.7.1 Delta	128
	6.7.2 Theta	132
	6.7.3 Gamma	134
	6.7.4 Vega	137
6.8	OTHER ASSETS	139
	6.8.1 Black-Scholes-Merton with Dividends	140
	6.8.2 Black-Scholes-Merton for Foreign-Exchange	140
	6.8.3 Black-scholes-Merton for Futures	141
6.9	DRAWBACKS OF THE BLACK-SCHOLES-MERTON MODEL	141
6.10	CHAPTER'S DIGEST	143
6.11	EXERCISES	143

Chapter 7 ▪ Monte Carlo Methods   147

7.1	THE NEED OF GENERAL OPTION PRICING TOOLS	147
7.2	MATHEMATICAL FOUNDATIONS OF MONTE CARLO METHODS	148
	7.2.1 Sample Means as Estimators of Theoretical Expectations	150
	7.2.2 The Laws of Large Numbers	151
	7.2.3 The Central Limit Theorem	154

7.3	OPTION PRICING WITH MONTE CARLO METHODS		155
	7.3.1	European Options that Depend Only on the Final Value of the Asset	156
	7.3.2	European Options that Depend on the Path of Asset Prices	158
7.4	EUROPEAN OPTIONS THAT DEPEND ON THE FINAL PRICE OF TWO ASSETS		162
7.5	CHAPTER'S DIGEST		165
7.6	EXERCISES		165

CHAPTER 8 ▪ The Volatility 169

8.1	HISTORICAL VOLATILITIES	169
8.2	THE SPOT VOLATILITY	171
8.3	THE IMPLIED VOLATILITY	172
8.4	CHAPTER'S DIGEST	174
8.5	EXERCISES	175

CHAPTER 9 ▪ Replicating Portfolios 177

9.1	REPLICATING PORTFOLIOS FOR THE BINOMIAL MODEL	177
9.2	REPLICATING PORTFOLIOS FOR THE BLACK-SCHOLES-MERTON MODE	181
9.3	CHAPTER'S DIGEST	187
9.4	EXERCISES	188

APPENDIX A ▪ Introduction to Python 191

A.1	BASIC OPERATIONS	191
A.2	DATA TYPES	192
A.3	VARIABLES	193
A.4	PRINT	194
A.5	PACKAGES	195

A.6		ROCKING LIKE A DATA SCIENTIST	195
	A.6.1	Import Data	196
	A.6.2	Using Dataframes	197
	A.6.3	Make Plot	201
A.7		CHAPTER'S DIGEST	206
A.8		EXERCISES	207

Appendix	B ▪ Introduction to Coding in Python	209
B.1	DEFINE YOUR OWN FUNCTIONS	209
B.2	IF	211
B.3	FOR	215
B.4	CREATING MATRICES	217
B.5	CHAPTER'S DIGEST	222
B.6	EXERCISES	223

Bibliography	225
Index	227

List of Figures

1.1	Future value with continuous compound interest rates	3
1.2	Present value with continuous compound interest rates	4
1.3	Coherence between the Future and Present value	5
2.1	Long and short forward payoff	10
2.2	Evolution comparison between an asset and its future	16
3.1	Long and short call/put options	26
3.2	Evolution comparison of ATM Call and Put options against the asset price	30
3.3	Call-Put Parity	33
3.4	Bull spread	35
3.5	Bear Spread	36
4.1	Binary Options	46
4.2	Forward start option	47
4.3	Compound option	48
4.4	Barrier Knock-in option	49
4.5	Barrier Knock-out	49
4.6	Lookback option	51
4.7	Asian options	52
5.1	Simple Binomial Model	56
5.2	General Binomial Model	56
5.3	Call option Payoff in the Binomial Model	57
5.4	General Call option Payoff in the Binomial Model	57
5.5	Put option Payoff in the Binomial Model	58

xiv ■ List of Figures

5.6	General Put option Payoff in the Binomial Model	58
5.7	Relationship between the asset price and the option price	59
5.8	Replication of Call option	61
5.9	Multi-Period Binomial Model	69
5.10	Labeling the scenarios of a Multi-Period Binomial model	70
5.11	How many times a scenario is reached depending on the time steps	72
5.12	Multi-Period Binomial Model. Simplifying the notation	74
5.13	Populating the tree prices	75
5.14	Finding the payoff	76
5.15	Calculating backwards the first layer	76
5.16	Calculating backwards the second layer	77
5.17	Populating the tree prices with a FX asset	79
5.18	Finding the payoff with a FX asset	79
5.19	Calculating backwards the first layer with a FX asset	80
5.20	Populating the tree prices when the asset pays dividends	82
5.21	Finding the payoff when the asset pays dividends	82
5.22	Calculating backwards the first layer the asset pays dividends	83
5.23	Considering the American option	85
5.24	Considering the American option when the asset pays dividends	86
5.25	Visualize the derivative concept	88
5.26	Taylor approximation of the exponential	89
5.27	Changing the shape	94
5.28	How the always raise scenario fills the matrix	97
5.29	How the payoff scenarios fills the matrix	97
5.30	Algorithm for the Backwards calculation	101
6.1	Price and Log-return of an asset	110
6.2	Histogram with Normal Distribution Curve	111
6.3	Model intuition	112
6.4	Accumulating days intuition	112

6.5	Black-Scholes-Merton formula for different maturities	119
6.6	Black-Scholes-Merton formula in 3D	120
6.7	Probability Distribution and the density function	124
6.8	ATM Delta of a call as a function of T	127
6.9	ATM Delta of a call as a function of T	130
6.10	In-the-money Delta ($S_0 = 110$) of a call as a function of T	130
6.11	Out-of-the-money Delta ($S_0 = 90$) of a call as a function of T	131
6.12	Call option delta surface	132
6.13	Theta for a put as a function of S_0	134
6.14	Put option theta surface	135
6.15	Gamma as a function of the Strike	136
6.16	Gamma as a function of time to maturity	136
6.17	Option's Gamma	137
6.18	Vega as a function of time to maturity	138
6.19	Option's vega	139
6.20	Comparison Fat-Tail vs Thin Tail	142
6.21	Volatility Clustering	142
7.1	Almost sure converge	152
7.2	Convergence in probability	153
7.3	Distribution of the empirical mean of 10.000 Binomial	155
8.1	Estimated daily volatility (in %) for the EURO STOXX50	171
8.2	Implied volatility surface	174
9.1	Binomial Model: asset prices	179
9.2	Binomial Model in Example 5.7	180
9.3	Comparison Option evolution with its replication portfolio rebalancing 10 times in a year	188
9.4	Comparison Option evolution with its replication portfolio rebalancing 100 times in a year	188

A.1	Dataframe	198
A.2	Dataframe description	199
A.3	Add a column	200
A.4	My first plot	201
A.5	Changing the size of a plot	202
A.6	Plot two time series	203
A.7	How subplot works	204
A.8	Two plots in one figure	205
A.9	A Histogram	206
B.1	Definition structure	210
B.2	Basic If Statement	212
B.3	If-Else Statement	213
B.4	If-Elif-Else Statement	214
B.5	For Statement	216

Foreword

The importance of derivatives in finance is beyond dispute. Its use ranges from speculation to risk management, and its users include not only banks and hedge funds but also retail investors, institutional investors, and insurance companies. Understanding the characteristics of derivatives is thus essential not only for those aspiring to a career at a bank or hedge fund, but also at asset managers, insurance firms, and even for private investors.

However, financial derivatives are a particularly difficult area for newcomers to gain an understanding of and comfort with. Not only is its mathematical framework daunting, but the world of financial derivatives also has its own jargon which may sound like an alien language when first heard. In addition, the derivatives landscape, be it from a regulatory, product, or technological point of view, is constantly shifting and changing.

As the word implies, a derivative is derived from the value of something else, called the underlying asset. The number of underlying financial assets, or the asset universe, is large and includes familiar assets such as equities, interest rates, foreign exchange, and commodities. But also, less familiar 'assets' such as mortality and longevity, and more recently digital coins. It is therefore unsurprising that traders and quants active in the field of derivatives are specialists in derivatives on one type of asset class or even in one type of derivative within a given asset class.

The reason for this is that there is not one mathematical model for all asset classes and derivatives. Each asset class and each derivative have its own features and may require product specific modeling. Nevertheless, the basic principle and single most important tenet of derivatives pricing, called the principle of no arbitrage, is the same across all asset classes. Yet, the principle of no arbitrage itself, when formulated in rigorous mathematical fashion, requires understanding of some deep results from measure theory. This is not a reason to despair as it is possible to understand and apply the statement of the far-reaching principle of no arbitrage without necessarily understanding its proof.

Despite the rather steep learning curve and hurdles to take to become familiar and comfortable with financial derivatives, the rewards are multitude. For me, the greatest reward has been the opportunity to learn from and to work with some of the best from both academia and industry, such as the authors of this book Elisa Alòs and Raúl Merino. It is thus not only a delight, but also an honor to have been given the opportunity to write this foreword to their book. Within the derivatives community, Elisa Alòs hardly needs an introduction. Her work on applications of Malliavin calculus to rough volatility models, and in particular her option decomposition formula, has shed light on many important aspects of options pricing under rough volatility models. I was therefore excited to hear about and read this book, which is a product of her collaboration with Raúl Merino, a senior quant with extensive industry experience who extended the decomposition formula in his doctoral dissertation.

When I first started out in quantitative finance some fifteen years ago, one of the challenges I faced was to find a text that explained the very basics of financial derivatives at a gentle pace without neglecting the mathematics underlying them. Learning a new subject is fraught with pitfalls. A solid grounding in the basics is of the utmost importance. This is perhaps even more so in finance, where, if you sit at a trading desk as a quant, there is little room for error. Skipping or not understanding the basics will without doubt lead to errors and therefore loss of valuable time and potentially a lot of money.

This book is the kind of book I wish I had access to when I first started to study derivatives. It covers the essentials, and it does it very well without assuming mathematical knowledge on the part of the reader beyond first-year calculus. The text provides a thorough introduction to the terminology and features of the most common derivatives, and even some exotic derivatives as well. The pricing of derivatives is first explained extensively within the framework of the simple binomial option pricing model. This is from a didactic point of view, that is the most logical choice, as the benchmark Black-Scholes-Merton model, also discussed in the book, can be regarded as the continuous limit of the binomial model. The latter chapters focus, among others, on the concept of a replicating portfolio, which is intimately related to the principle of no arbitrage I mentioned earlier.

However, the text does not distinguish itself from others only on the grounds of lucid explanations of the theoretical basics of derivatives pricing, but also by the numerical examples and code scattered throughout the text. At every stage and chapter, the reader will find engaging numerical demonstrations and exercises that further clarify the theory behind derivatives pricing. Furthermore, the numerical examples are coded in Python, which is now one of the most widely used programming languages at banks and asset managers. After reading the next two-hundred or so pages, the student will have gained the necessary theoretical background as well as hands-on programming experience, which is an indispensable skill in quantitative finance.

Without further ado, I wish you, the reader, an exciting journey into the fascinating world of financial derivatives. With this book, you shall be in good hands!

Frido Rolloos

Preface

> *He who would learn to fly*
> *one day must first learn to stand*
> *and walk and run and climb and dance;*
> *one cannot fly into flying.*
>
> **Friedrich Nietzsche**
> **Thus Spoke Zarathustra, 1883**

Since the inception of the Chicago Board of Options Exchange, the use of derivatives has been unstoppable. On the one hand, electronic trading and computers have enabled us the possibility to calculate and trade in seconds. As well as receive and process a massive amount of data in seconds. On the other hand, there has been extensive development in a wide variety of areas. From the product side, creating new derivatives types, to the hectic research covering theory, models, and computational methods, among others.

Financial instruments such as forwards, options, and swaps are used intensively in the financial industry. Although they can be used to invest or speculate, they are essential for risk management, allowing investors to reduce their exposure to a specific risk. These products are called *derivatives* because their value depends on the evolution of another asset, for example, a stock, an index, an interest rate, a commodity, etc.

The main problems related to derivatives are *pricing* and *hedging*. *Pricing* means determining the derivative price, which is difficult because it depends on the evolution of another asset, the underlying. Whereas *hedging* is connected to risk management, since the aim is to reduce risk exposure. It is possible to do that using a hedging strategy based on the portfolio construction that behaves similarly to the derivative.

The above problems are not easy to address. Intuitively, to obtain a derivative price, we think about forecasting the evolution of the asset. But if not done carefully, that approach creates arbitrage. Arbitrage means, roughly speaking, 'making money from nothing'. We consider the market efficient, or smart enough, not to allow arbitrage opportunities. In other words, all the market quotes must be consistent with each other.

The above non-arbitrage price is not easy to establish. It is a demanding task from a mathematical and computational point of view. Concepts and tools such as non-arbitrage prices, binomial trees, implied volatilities, Monte Carlo methods, programming, or computational costs are the basis of daily practice in the financial industry.

From an academic point of view, most finance and business schools have a compulsory course where derivatives are introduced. These courses have to focus not only on the theoretical aspects but also on practical issues. The purpose of this book is twofold. On the one hand, the book introduces the theory clearly. But, on the other hand, it focuses on the practical side. For this reason, we also focus on developing your own Python code. The book is not an introduction to programming. The aim is to show that with basic knowledge, it is possible to price your own derivatives and code all the theoretical tools explained.

This book is primarily addressed to undergraduate students in Economics and related areas. No prior programming knowledge or advanced mathematics is required. It is organized as follows. Chapters 1–4 focus on the main types of derivatives, such as futures, forwards, vanilla, and exotic options. Chapters 5–9 are devoted to mathematical tools such as binomial trees, the Black-Scholes-Merton model, Monte Carlo methods, implied volatilities, replication portfolios, and the calculation of the Greeks (the sensitivity of option prices to market parameters). In the Appendix, we include an introduction to Python.

Finally, the book contains a Github repository (accessible here: https://bit.ly/3bllnuf) with Python notebooks ordered by chapter. The notebooks do not claim to be industrial or efficient, nor follow a standard like PEP8. The purpose is to support the theory and broaden the reader's perspective because mathematical finance is not just about theoretical formulas, but also about putting them into practice. At best, it will motivate the reader to learn more about coding.

Aknowledgements We are deeply indebted to all our students of the course *Financial Derivatives and Risk Management* at the Universitat Pompeu Fabra. They have allowed us to improve, year after year, our teaching and have inspired this book. Moreover, we would like to thank all our friends and colleagues who work every day in the effective construction of bridges between industry and academia and who gave us fruitful advice that improved this book. Specially, we want to thank David García-Lorite, who has been there when we needed his suggestions and comments; Suren Harutyunyan for his detailed review of the text; and Frido Rolloos, who has been a strong supporter and a source of fruitful discussions. We would also like to thank our editors Callum Fraser and Mansi Kabra for their help and support. All remaining errors are ours.

Elisa Alòs and Raúl Merino
Barcelona, July 2022

CHAPTER 1

Introduction

1.1 FINANCIAL MARKETS

Finances are a natural element for people. We met together and exchanged resources since the beginning of time. In a financial market, people trade securities at a low transaction cost. The listed price reflects the asset's supply and demand. But the purpose is to transfer risks in exchange for investment opportunities.

We may trade financial instruments at central locations, such as organized exchanges. But they can also be negotiated in a private contract between two parties in an Over-The-Counter (OTC) contract. This makes it possible to design tailor-made agreements even when counterparty risk exposition increases.

We can execute a trade for different reasons, such as investing, hedging, or speculating.

- **Investing** is the act of distributing financial resources into assets to generate financial returns. It's a long-term initiative.

- **Hedging** is the use of financial instruments to mitigate or eliminate certain types of risks. Hedge strategies are designed to reduce risks, rather than to generate returns.

- **Speculating** is based on trading in the financial markets to generate short-term gains. The trader assumes the risk of losing money but holds the expectation of obtaining a significant profit.

1.2 DERIVATIVES

Financial market instruments can be divided into two different categories. On the one hand, we have the 'prime source' assets, which we will

DOI: 10.1201/9781003266730-1

refer to as 'underlyings', and which can be stocks, bonds, commodities, foreign currencies, etc. On the other hand, their 'derivative' contracts, financial claims that promise some payment or delivery in the future, depending on the behavior of the underlying.

The most typical financial derivatives are futures (or forwards) and options. A future (or forward) is a legal agreement to buy or sell a particular asset at a predetermined price and at a specified time in the future. Meanwhile, an option contract gives the right but not the obligation to buy (or sell) a particular asset at a predetermined price and at a specified time in the future.

Many people think that derivative contracts, such as futures and options, are inventions of the modern economy. However, derivative contracts emerged as soon as humans could make credible promises. They were the first instruments to guarantee the supply of basic products, facilitate trade and insure farmers against the loss of crops. The first written evidence of a derivative contract was in law 48 of the Hammurabi code, roughly between 1782 to 1750 BCE.

One of the first stories related to the speculation of derivatives is due to Thales of Mileto. Thales made a deposit at the local olive presses. As nobody knew for sure whether the harvest would be good or bad, Thales purchased the rights to the presses at a relatively low rate. When the harvest proved to be abundant, the demand for the presses was high, Thales charged a high price for their use and reaped a considerable profit.

The earliest evidence of organized exchange was in the 1730s, when the Tokugawa shogunate authorized rice futures trading. After that, the Dojima Rice Exchange opened. Nearly a century later, in 1848, the Chicago Board Of Trade (CBOT) was founded as an association of grain merchants, see (Algieri, 2018). The first applications of derivatives were related to the use of commodities.

In 1973, the world's first listed options exchange, the Chicago Board of Options Exchange (CBOE), opened in Chicago. The same year that the Black-Scholes-Merton formula was published. Since then, the trading volume of derivatives has increased largely.

1.3 TIME HAS A VALUE

A natural process in finance is to lend and borrow money. The question arises of how much we should ask in exchange for lending money for a fixed term. Although the borrower returns the same amount to us, it does not have the same value. After all, we have an opportunity cost since we could have made a return in that period. Not to mention other risks. In other words, 1€ today does not have the same value as 1€ tomorrow. Therefore, money has a time value.

An important tool in financial mathematics is to calculate how much capital we have to receive/pay at a future date when lending/borrowing. There are three different methodologies to calculate it. If P is the initial capital, r is the interest rate, and t is the term, the capital to be returned for each method is as follows:

- **Simple interest:** $P(1+rt)$.

- **Continuous interest** $P(1+r)^t$.

- **Continuous compound:** Pe^{rT}.

The amount of money we have to return at a future time reflects the value of today's money with an interest rate of r. So, we will refer to it as the future value or FV (see Figure 1.1). The interest methodology usually depends on the asset, the term, and the contract.

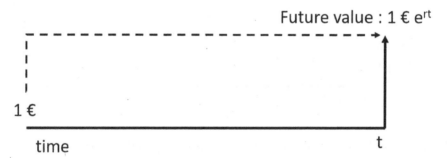

Figure 1.1 Future value with continuous compound interest rates.

Example 1.1

Calculate the value of 1€ at 30 years with an interest rate of 10% using:

1. Simple interest rates.

2. Compound interest rates.

3. Continuously compound interest rates.

1. 1€ $(1 + 10\% \cdot 30) = 4$ €.
2. 1€ $(1 + 10\%)^{30} = 17.45$ €.
3. 1€ $e^{10\% 30} = 20.085$ €.

Another important tool is learning to bring future cash flows to today. For example, if someone is going to pay us a stream of cash flows in the future, what would the present value be? For doing that, we calculate the Present Value. We will refer to it as PV (see Figure 1.2). In this case, we have three possibilities as well.

- **Simple interest**: $PV(t) = \frac{P}{1+rt}$.
- **Continuous interest**: $PV(t) = P(1+r)^{-t}$.
- **Continuous compound**: $PV(t) = Pe^{-rT}$.

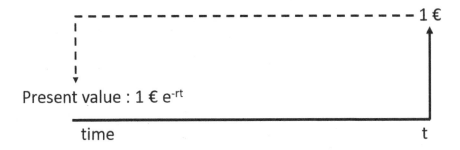

Figure 1.2 Present value with continuous compound interest rates.

Example 1.2

Assume that the interest rates are 5% for 1 year and 6% for 2 years.

1. How much will worth today earn 100€ in 1 year using continuous interests?
2. How much will worth today earn 100€ in 2 years using continuous interests?

1. 100 € $(1 + 5\%)^{-1} = 95.23$€ .
2. 100 € $(1 + 6\%)^{-2} = 88.99$€ .

There is a principle of coherence. When using a methodology to calculate the future value, we have to use the same methodology to calculate the present value. It must be consistent. That means that when moving an initial capital to the future and bringing it back to today, we must obtain the same initial capital (see Figure 1.3). In mathematical terms, we have the equality:

$$FV(t) \cdot PV(t) = 1.$$

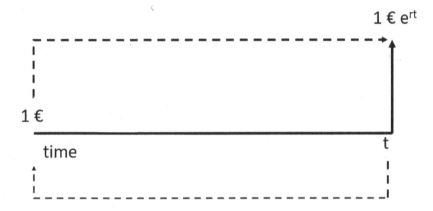

Figure 1.3 Coherence between the Future and Present values.

In all the book, for simplicity, we will use continuous compound interest rates.

1.4 NO-ARBITRAGE PRINCIPLE

When we calculate theoretical prices, there is a principle that we always assume. There is no arbitrage. We say there is an arbitrage when we can obtain a strictly positive benefit without risk. In other words, it is impossible to find a 5€ bill on the street or to have a free lunch.

In the market, it is possible to find arbitrages. But they are difficult to exploit. Their profit has to be greater than the trading cost, and they disappear quickly from the market.

By using this condition, we have the following consequence. There are two portfolio A_t and B_t. If these portfolios have the same value at a future time T, i.e. $A_T = B_T$, then it must be worth the same today, i.e. $A_0 = B_0$. We can prove this result by using the reduction to the absurd.

6 ■ Introduction

First, we assume that $A_0 > B_0$. Then, we create a strategy where:

- We buy the portfolio B_0.
- We sell the portfolio A_0.

At the inception time, we have the profit $A_0 - B_0$. Then, at T:

- We have the portfolio B_T.
- We owe the portfolio A_T.
- We have $(A_0 - B_0)e^{rT}$ units in a cash account, since we deposited the profit.

Therefore, at T our strategy has a value of
$$B_T - A_T + (A_0 - B_0)e^{rT},$$
but $A_T = B_T$ by hypothesis. Then, we have a profit of
$$(A_0 - B_0)e^{rT} > 0.$$
So, there is an arbitrage which we assumed was not possible.

Now, we have to prove the alternative case. We assume that $B_0 > A_0$. Then, we create a strategy where:

- We buy the portfolio A_0.
- We sell the portfolio B_0.

At the inception time, we have the profit $B_0 - A_0$. Then, at T:

- We have the portfolio A_T.
- We owe the portfolio B_T.
- We have $(B_0 - A_0)e^{rT}$ units in a cash account, since we deposited the profit.

Therefore, at T our strategy has a value of
$$A_T - B_T + (B_0 - A_0)e^{rT},$$
but $A_T = B_T$ by hypothesis. Then, we have a profit of
$$(B_0 - A_0)e^{rT} > 0.$$
So, there is an arbitrage which we assumed was not possible. If it is impossible that $A_0 > B_0$ and $B_0 > A_0$, then it must be $B_0 = A_0$.

1.5 CHAPTER'S DIGEST

In this introduction, we explained that the financial market helps us to exchange risk for investment opportunities. The trades can be executed in an organized exchange or between two counterparts. There are various reasons to trade like investing, hedging, or speculating.

Later, we explain what a derivative is and the most popular ones: forwards and options. We reviewed the first historical uses from Hammurabi's code to the Chicago Board of Options Exchange, including how Thales of Mileto became wealthy trading futures or the first organized exchange in Japan. There is a common factor, all the historical uses of derivatives are related to commodities.

Then we do a quick review of the future and present value calculations. There are three possible methodologies to calculate it: simple, compound, and continuous compound interest. In the book, we always use continuous compound interest, but in real life, we need to check the contract where it is specified. It can depend on the asset, product, and time to maturity. In the case of derivatives, it is useful to calculate the no-arbitrage price. It is important to remember that time is valuable as we can lose investment opportunities.

Finally, we explain what an arbitrage is. We see that if two portfolios have the same future value, or payoff, then they must be worth the same. It is proved mathematically by reduction to absurd.

1.6 EXERCISES

1. Hedging is used to:

    (a) Reduce risk.

    (b) Speculation.

    (c) Increase exposure to price movements.

2. If two financial products, A and B, have the same payoff at T. At $t < T$,

    (a) They should worth the same.

    (b) It is not possible to know anything at t.

    (c) $A > B$.

3. Suppose 1.000€ is deposited for 6 years in an account paying 4.25% per year compounded annually.

    (a) Find the future value. **Solution: 1,283.68€**

    (b) Find the amount of interest earned. **Solution: 283.68€**

4. If you invest 1,000,000€ in an investment paying a 2% compound continuously, how much will you have after 5 years? **Solution: 1,105,170.92€**

5. Which profit I will obtain if I lend 50,000€ on a start up during 5 months at 7% compound continuously? **Solution: 1,479.81€**

6. If I have to pay 500€ in 3 years and the interest rates are 2%, how much this is worth it today with continuous compound rates? **Solution: 470.88€**

CHAPTER 2

Futures and Forwards

Forward and futures contracts are the simplest examples of derivatives. In this chapter, we review the basic concepts and tools related to these financial instruments.

2.1 FORWARD CONTRACTS: DEFINITIONS

A **forward contract** is an agreement between two parties where one of them agrees to buy (and the other one agrees to sell) an underlying asset (a stock, an index, a commodity, etc.) at some future time T (that we call the **maturity time**) and at a certain fixed delivery price F. The party that agrees to buy is said to take a **long position**, whereas the party who agrees to sell takes a **short position**. Notice that the delivery can be **physical** or **in cash**. In this latest case, the party in the long forward does not receive the underlying asset, but simply its value at maturity.

If we denote by S_T the value of the underlying asset at maturity, the final value of the long-forward contract (its **payoff**) is equal to

$$S_T - F,$$

and the payoff of the short-forward contract is given by

$$F - S_T,$$

as we can see in Figure 2.1.

DOI: 10.1201/9781003266730-2

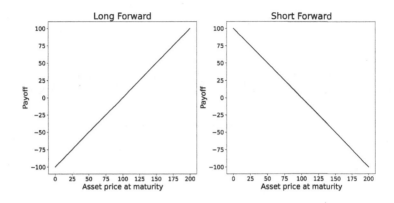

Figure 2.1 Long- and short-foward payoff with delivery price $F = 100$, in terms of the underlying asset price at maturity.

If the final asset price S_T is over F, the long position gets a positive payoff while the short position gets a loss. On the contrary, if S_T falls below F, the payoff of the long forward is positive, while the payoff of the short forward is negative.

Notice that we can write the payoff of the long forward as $h(S_T)$, where

$$h(x) := x - F.$$

The function h is called the **payoff function** of the long forward. In a similar way, the payoff function h' for the short-forward contract is given by

$$h'(x) = F - x.$$

Many other derivatives, as we will see in the following chapters, are defined by their payoff functions.

Remark 2.1.1 *Notice that (assuming that asset prices are positive) the payoff of a long-forward contract is bounded below by $-F$, and then the losses can not be bigger than F. But the payoff of a short forward is unbounded below, so there is no limit for potential looses.*

The delivery price F is determined in such a way that the forward contract is worth zero at inception from both sides. That is, in order the contract to be 'fair'. It is tempting to assume that this delivery price has to be the expected value of the underlying asset. But this is not true. The 'fair' value of the delivery price is the non-arbitrage price, that is, the value of F that does not allow for arbitrage opportunities. This non-arbitrage price is called the **forward price** and it depends on today's price on the underlying asset, interest rates, and time to maturity, as we see in detail in Section 2.4. As this 'fair' price depends on today's data and on time to maturity, it changes from day to day. Then, from now on, we will denote by F_t the forward price computed at time t. Notice that, as we approach maturity, F_t has to tend to the spot price. Otherwise, arbitrage opportunities appear, because we can buy/sell an asset at a price that is less/higher than its market price. For example, if F_t is clearly cheaper than the asset price just before maturity, one could buy the asset at the price F_t, selling latter the asset at its market price, leading to an arbitrage.

2.2 FUTURES

A **futures contract** is a derivative with the same payoff as a forward contract, but we commonly use the word 'forward' when the contract is traded OTC, and the word 'future' when it is traded through an exchange. The fair price of a futures contract is called the **future price**, that is, computed in the same way as the forward price. That is, forward and futures contracts are essentially equal from the mathematical point of view, but they are traded in a different way.

In a futures contract, settlement is usually made in cash. Moreover, most futures contracts are closed out before maturity by simply taking the opposite position. One big difference with the mechanism of forward contracts is the **daily settlement**. In a forward contract no money changes hands before maturity. But in a futures contract, money moves to one party to the other each day, according to the evolution of the corresponding future price. Let us see how this works. To fix ideas, let us assume that we have a long future with payoff

$$S_T - F_t.$$

As the future price gets equal to the asset price as we approach maturity, we have that $S_T = F_T$ and then we can write the above expression as

$$F_T - F_t.$$

Now imagine that the difference between T and t is equal to n trading days. Then, if we denote

$$t_i := t + i\frac{T-t}{n}, \quad i = 1, ..., n$$

the days before 'today' and maturity (notice that $t_0 = t$ and $t_n = T$), we can write

$$\begin{aligned}F_T - F_t = F_T &- F_{t_{n-1}} \\ &+ F_{t_{n-1}} - F_{t_{n-2}} \\ & F_{t_{n-2}} - \cdots - F_{t_2} \\ &\phantom{+F_{t_{n-2}} - \cdots} + F_{t_2} - F_{t_1} \\ &\phantom{+F_{t_{n-2}} - \cdots - F_{t_2}} + F_{t_1} - F_t.\end{aligned}$$

That is,

$$F_T - F_t = \sum_{i=0}^{n}(F_{t_{i+1}} - F_{t_i}),$$

where the sumands $F_{t_{i+1}} - F_{t_i}$, $i = 0, ..., n$ are the daily variations of the future price. Then, instead of doing the payment at maturity, we pay every day i the variation $F_{t_{i+1}} - F_{t_i}$ of the future price.

In order to manage these payments, the exchanges require depositing an **initial margin**, an initial amount of money, in a **margin account**. As the future price changes, money is withdrawn (if $F_{t_{i+1}} - F_{t_i}$ is positive) or deposited (if $F_{t_{i+1}} - F_{t_i}$ is negative) from it. The balance of the margin account must stay above a level called **maintenance margin**. If it ever falls below this quantity, the holder of the contract receives a **margin call** and is required to deposit more money in order for the balance to be equal again to the initial margin.

Let us see with an example how this works.

Example 2.1

Assume that we go long in a futures contract with maturity in 5 days, with today's future price equal to 100€. The initial margin is assumed to be 15€ and the maintenance level equal to 10€. Now, assume that the evolution of the future price in the next 5 days is (in €):

$$101, 104, 89, 90, 90.$$

Then the evolution of the balance of the margin account is as follows:

- The first day, the increment of the future price is given by 101€ − 100€ = 1€. Now 1€ is deposited in the margin account, which is now worth 16€.

- The second day, the increment of the future price is equal to 104€ − 101€ = 3€. Then, 3€ are deposited in the margin account, and its value is now 19€.

- The third day, the increment of the future price is 89€ − 104€ = − 15€. Now 15€ are withdrawn from the margin account, whose value is now 4€. As this value is less than the maintenance level, we receive a margin call, asking us to deposit 15€ − 4€=11€ tomorrow.

- The fourth day, we have to deposit 11€ as required in the margin call. On the other side, the increment in the future price is equal to 90€ − 89€ = 1€. Then, the new balance of the margin account is 4€+11€+1€=16€.

- The last day, the increment in the future price is equal to zero, and the final balance of the margin account is equal to 16€. Then, the final benefit is the difference between the final balance of this margin account (16€) and the quantity we deposited in the margin account (15€+11€=26€). That is, −10 €. Notice that this quantity is equal to the payoff $S_T - 100 = 90 - 100$ of the future contract.

Similarly, if we have a short-forward contract with payoff

$$F_t - S_T = F_t - F_T$$

we can write

$$F_t - F_T = \sum_{i=0}^{n} -(F_{t_{i+1}} - F_{t_i}),$$

where the sumands $F_{t_{i+1}} - F_{t_i}$, $i = 0, ..., n$ are again the daily variations of the future price. Then, instead of doing the payment at maturity, we pay every day i the variation $F_{t_i} - F_{t_{i+1}}$.

Let us see how this works, by analyzing the short position of the same future contract as in Example 2.2.

Example 2.2

Assume that we go short on a futures contract with a maturity in 5 days, with today's future price equal to 100€. The initial margin is assumed to be 15€ and the maintenance level equal to 10€. Now, assume that the evolution of the future price in the next 10 days is

(in €):
$$101, 104, 89, 90, 90.$$

Then the evolution of the balance of the margin account is as follows:

- The first day, the increment of the future price is given by 101€ − 100€ = 1€. Now 1€ is withdrawn from the margin account, which is now worth 14€.

- The second day, the increment of the future price is equal to 104 € − 101€ = 3€. Then, 3€ are withdrawn from the margin account, and its value is now 11€.

- The third day, the increment of the future price is 89€ − 104€ = 15€. Now 15€ are deposited in the margin account, whose value is now 26€.

- The fourth day, the increment in the future price is equal to 90€ − 89€ = 1€. Then, the new balance of the margin account is 26 − 1€ = 25€.

- The last day, the increment in the future price is equal to zero, and the final balance of the margin account is equal to 25€. Then, the final benefit is the difference between the final balance of this margin account (25€) and the quantity we deposited in the margin account (15€). That is, 10 €. Notice that this quantity is equal to the payoff $100 - S_T = 100 - 90$ of the future contract.

2.3 WHY TO USE FORWARDS AND FUTURES?

Historically, forwards and futures appear to lock the price of commodities. So, their first application is hedging. Nowadays, futures and forwards play a relevant role in risk management, allowing us to fix a future price not only for commodities, but also, for example, for other financial instruments such as currencies.

Nevertheless, these financial instruments are very attractive for speculators because the initial price we have to pay to enter into these contracts is equal to zero. The counterpart is the risk. Notice that the final payoff of the long forward contract is $S_T - F$, a quantity that can be negative if asset prices fall. For the short contract, the risk is higher: its final payoff is given by $F - S_T$, a quantity that is an unbounded function of the final asset price S_T. That is, the payoff can be not only negative, but very big in absolute value if the asset price rises. In fact, the quantity you can lose is unlimited.

2.4 THE FAIR DELIVERY PRICE: THE FORWARD PRICE

Given the underlying asset and the maturity time T, we have said that the delivery price F is computed in such a way that the contract is worth zero at inception for both parties. That is, that the contract is 'fair'. It is tempting to think that F has to be the expected value of the asset price at maturity. But this is not true. As introduced in Chapter 1, F has be computed in such a way there are no arbitrage opportunities. This non-arbitrage value of F is called the **forward/future price**.

2.4.1 The General Approach

Let us see the computation of the fair value for the delivery price F in the simplest case of a non-paying dividend stock (other cases are studied in Section 2.4.2). According to the arguments in Chapter 1, the value of a long forward with a payoff $S_T - F$ is the same as the value of a portfolio composed of stocks and a risk-free investment with the same value at maturity. How can we construct such a portfolio? Consider the simplest case when the underlying is a non-paying dividend stock and assume we are at a certain moment $t < T$. Then one can simply take

- A stock, with today's value S_t, and final value of S_T, and

- a risk-free investment equal to the present value of F, that is, equal to $Fe^{-r(T-t)}$.

At T, the portfolio's value is $S_T - F$, the same that the long forward. But today's value of this portfolio is equal to $S_t - Fe^{-r(T-t)}$. As this value is equal to today's value of the long forward (that is worth zero), one gets $S_t = Fe^{-r(T-t)}$, which implies

$$F = S_t e^{r(T-t)}. \tag{2.1}$$

As this value depends on the time moment t, from now on we will use the notation F_t to refer to the forward price computed at time t.

Remark 2.4.1 *Notice that this price has been obtained simply by using non-arbitrage arguments. We have not computed expectations, nor we have considered a specific model.*

The application of this formula is straightforward, as we can see in the following example

---
Example 2.3

Suppose that you enter into a 6-month forward contract on a non-dividend paying stock when the stock price is 100€ and the annual risk-free interest rate (with continuous compounding) is 10%. What is the forward price?

As $S_t = 100$, $r = 0.1$, and time to maturity is equal to half a year (that is, $T - t = 0.5$), a direct application of Equation (2.1) gives us that the forward price is given by

$$100e^{0.10 \times 0.5} = 105.1271.$$

Remark 2.4.2 *Observe that the difference between the asset price S_t and the forward price F_t is equal to*

$$S_t - F_t = S_t - S_t e^{r(T-t)} = S_t(1 - e^{r(T-t)}),$$

which tends to zero as t approaches T (see Figure 2.2).

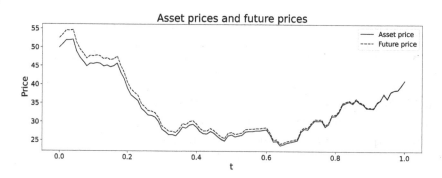

Figure 2.2 Simulation of asset prices and the corresponding future prices[2].

---

[2] Asset prices have been simulated assuming a Black-Scholes-Merton model with $S_0 = 100, \mu = 0.05$, and $\sigma = 0.3$. We will study this model in Section 6.

The Fair Delivery Price: The Forward Price

This forward price F_t is really the only non-arbitrage delivery price. Imagine, for example, that the delivery price F is greater than the forward price $F_t = S_t e^{-r(T-f)}$. Then we can create an arbitrage strategy by taking:

- A short forward, whose value today is 0, and with a final payoff $F - S_T$.
- A stock, with today's value equal to S_t and final value S_T.
- A risk-free investment is equal to $-S_t$ (that is, we borrow money), with a final value $-S_t e^{r(T-t)}$.

Today's value of this strategy is zero, while its final value is $F - S_t e^{r(T-t)} > 0$. That is, we have constructed an arbitrage strategy. In a similar way, if $F < S_t e^{rT}$ we can take:

- A long forward, with today's value equal to 0 and with final payoff $S_T - F$.
- A short stock (that is, we short sell a stock), with today's value equal to S_t and final value $-S_T$.
- A risk-free investment equal to S_t, with a value at maturity equal to $S_t e^{r(T-t)}$.

This portfolio is worth zero at inception and has a final value equal to $S_t e^{r(T-t)} - F > 0$.

2.4.2 Some Special Cases

In some cases, the formula for the forward price we have developed is not the right one due to the fact that the underlying asset provides an income (that can be positive or negative). Classical examples of assets providing a positive income include paying-dividend stocks. On the other hand, several commodities provide a negative income due to storage costs. Storage costs play a relevant role in trading commodities. Notice, for example, that during the COVID-19 crisis, in April 2020, oil prices got to negative values.

Let us see some examples of these scenarios.

2.4.2.1 Assets that Provide a Known Income

In some cases, the investment asset provides a known income I at some moment $s \leq T$. We notice that this income can be positive (as in the case of paying-dividend stocks) or negative (as, for example, in the case of storage costs).

The main difference with the scenario in Section 2.4.1 is that the holder of the asset has a wealth that is not equal to S_T but equal to $S_T + Ie^{r(T-s)}$. That is, equal to the final value of the asset plus the future value of the income.

This implies that, if we want to construct a portfolio with the same payoff as a long forward $S_T - F_t$ replicating the arguments in Section 2.4.1, we have to do adequate corrections. More precisely, we have to modify our risk-free investment to 'compensate' this extra final wealth $Ie^{r(T-s)}$. More precisely, we should take:

- One unit of the the underlying, and
- A risk free investment equal to $-F_t e^{-r(T-t)} - Ie^{-r(s-t)}$

Notice that, at maturity, the value of this portfolio is given by

$$S_T - F_t - Ie^{-r(s-t)}e^{r(T-t)} + Ie^{r(T-s)} = S_T - F_t,$$

as we desired. Now, similar non-arbitrage arguments as in Section 2.4.1 allow us to compute F_t. More precisely, today's value of the portfolio is given by

$$S_t - F_t e^{-r(T-t)} - Ie^{-r(s-t)}.$$

As the contract has to be worth zero today,

$$S_t - F_t e^{-r(T-t)} - Ie^{-r(s-t)} = 0,$$

this implies that

$$F_t = S_t e^{r(T-t)} - Ie^{r(T-s)}.$$

Notice that the above formula can be extended to the case that we receive more than one income before maturity. It is the case that we receive some income (as dividends) at some fixed dates, or that we have to do some payments at some moments. In this scenario, the forward price at time t is given by

$$F_t = S_t e^{r(T-t)} - \sum_{i=1}^{n} I_i e^{r(T-s_i)}, \qquad (2.2)$$

where $I_1, ..., I_n$ are the incomes received at the time moments $s_1, ..., s_n$, respectively. Let us see the application of the above formula in some examples.

Example 2.4

Consider a 5-month forward contract on a stock when the stock price is 100€. We assume that the risk-free rate of interest (continuously compounded) is 5% per annum for all maturities. We also assume that dividends of 0.20€ per share are expected after 2 months and 4 months.

In this case, time to maturity $T - t$ is equal to 5 months, that is, $\frac{5}{12}$ years. Denote now by s_1 and s_2 the moments we receive the first and the second dividend, respectively. As the first dividend is received in 2 months, this means that $T - s_1$ is equal to 3 months ($\frac{3}{12}$ years). In a similar way, $T - s_2$, is equal to $\frac{1}{12}$ years. Then the forward price is given by

$$100e^{0.05 \times \frac{5}{12}} - 0.2e^{0.05 \times \frac{3}{12}} - 0.2e^{0.05 \times \frac{1}{12}} = 101.7018.$$

Example 2.5

Consider a 2-year futures contract on an investment asset that provides no income. It costs 5€ per unit to store the asset, with the payment being made in 18 months. Assume that the spot price is 100€ per unit and the risk-free rate is 5% per annum (continuously compounded) for all maturities.

Because of the storage costs, we have a negative income of 5€ in 18 months. If we denote by s the moment we have to do the payment, we can see that $T - s$ is equal to 6 months, that is, 0.5 years. Then the forward price is given by

$$100e^{0.05 \times 2} + 5e^{0.05 \times 0.5} = 115.6437.$$

2.4.2.2 Assets that Provide an Income Proportional to Its Price

Another scenario is when the asset provides an income, but this income is not a fixed quantity, but it is proportional to the final value of the asset. A typical example is given by currencies. If the value of, for example, a yen is equal today to $S_t = 0.0077$ euros, its value at maturity will be equal to $S_T e^{c(T-t)}$, where S_T denotes the price of a yen at maturity, and c denotes the Japanese interest rate. Another example is the case when the underlying asset is an index. As the index can be seen as a basket of paying-dividend stocks, the index pays a dividend. Usually, for the sake of simplicity, this dividend is assumed to be a dividend yield, not a known income. Then the final value of the investment in the index is equal to $S_T e^{c(T-t)}$, where c denotes now the convenience yield.

In both cases, if we want to get a final payoff of $S_T - F_t$, we have to 'compensate' this extra income. Towards this end, we can construct a porfolio consisting of:

- $e^{-c(T-t)}$ assets, and

- a risk-free investment equal to $-F_t e^{-r(T-t)}$.

Notice that, at maturity, this portfolio is worth $e^{-c(T-t)} S_T e^{c(T-t)} - F_t = S_T - F_t$ as we desired. On the other hand, today's value of the portfolio is equal to
$$S_t e^{-c(T-t)} - F_t e^{-r(T-t)}.$$

As this quantity has to be worth zero,
$$F_t = S_t e^{(r-c)(T-t)}, \qquad (2.3)$$

which gives us the expression for the forward price in this scenario. Let us see some examples.

Example 2.6

Assume that today's price of the yen is 0.0077€ . If the Japanese interest rate is 1% (cc) and the European interest rate is 2% (cc), what is today's forward price of the yen, if maturity T is equal to 2 years?

As the foreign interest rate is $c = 0.01$, a direct application of 2.3 gives us that the forward price is
$$F_t = 0.0077 e^{(0.02-0.01) \times 2} = 0.007141409.$$

Example 2.7

Assume that today's value of an index is equal to 1000. It gives a dividend yield of 2% (cc) and the European interest rate is 1% (cc), what is today's forward price of the index if maturity T is equal to 1 year?

As the dividend yield is $c = 0.02$, a direct application of Equation (2.3) gives us that the forward price is

$$F_t = 1000e^{(0.01-0.02)\times 1} = 990.0498.$$

2.4.3 The Price of a Forward Contract

The forward contract is a financial instrument, and then it has a value. We have seen that the price of the forward contract is zero at inception t and equal to the corresponding payoff function at maturity. But, what is the value of the forward contract at any time moment u between the inception of the contract and its maturity time?

Again, the arguments in Chapter 1 tell us that the value of a long forward with payoff $S_T - F_t$ is the price of a portfolio composed of stocks and a risk-free investment with the same value at maturity. How can we construct such a portfolio? Let us see how to compute the price of a forward contract in different scenarios.

2.4.4 The general case

Consider again the simplest case when the underlying is a non-paying dividend stock and assume we are at a certain moment $u \in [t, T]$. Then one can simply take

- A stock, with today's value S_u, and final value of S_T, and
- a risk-free investment equal to the present value of F_t, that is, equal to $F_t e^{-r(T-u)}$.

Today's value of this portfolio is equal to

$$S_u - F_t e^{-r(T-u)},$$

which implies that this is the price of the long forward contract at any moment u before maturity. Notice that, if $u = T$, the above price reads as $S_T - F_t$, which coincides with the payoff of the long forward contract.

Remark 2.4.3 *The price of the short forward contract is given by*

$$F_t e^{-r(T-u)} - S_u,$$

due to the fact that this is the opposite position.

2.4.4.1 The Case of a Known Income

Now consider the case that the underlying asset generates some incomes $I_1, .., I_n$ at some moments $s_1, ..., s_n$. Then, in order to construct a payoff with final value $S_T - F_t$, one should take

- A stock, with today's value S_u, and final value of S_T, and
- a risk-free investment equal to

$$-F_t e^{-r(T-u)} - \sum_{s_i \geq u} I e^{-r(s_i-u)}.$$

Notice that the final value of this portfolio is, as desired,

$$S_T + \sum_{i=1}^{n} I_i e^{r(T-s_i)} - F_t - \sum_{s_i \geq u} I e^{-r(T-s_i)} = S_T - F_t.$$

while its value is given, at time u, by

$$S_u - F_t e^{-r(T-u)} - \sum_{s_i \geq u} I e^{-r(s_i-u)}.$$

2.4.4.2 Assets that Provide an Income Proportional to Its Price

Let us see how to construct in this case a portfolio at time $u \in (t, T)$, with final payoff $S_T - F_t$. Towards this end, we can construct a porfolio consisting of:

- $e^{-c(T-u)}$ assets, and
- a risk-free investment equal to $-F_t e^{-r(T-u)}$.

Notice that, at maturity, this portfolio is worth

$$S_T e^{c(T-u)} e^{-c(T-t)} - F_t e^{-r(T-u)} e^{-r(T-u)} = S_T - F_t$$

as we desired. On the other hand, the value of the portfolio at time u is equal to

$$e^{-c(T-u)} S_u - F_t e^{-r(T-u)}.$$

2.5 CHAPTER'S DIGEST

A **forward contract** is an agreement between two parties where one of them agrees to buy (and the other one agrees to sell) an underlying asset at some future time T (the **maturity time**) and at certain fixed delivery price F. The party that agrees to buy is said to take the **long position**, where the party who agrees to sell takes the **short position**. A **futures contract** is similar to a forward contract, but forwards are traded OTC, while futures are traded through an exchange. The mechanisms of these contracts are also different. In a forward contract, no money exchange hands before maturity, while in a futures contract, the settlement is done daily. That is, each day, the party that loses pays to the party that gets a benefit.

The delivery price F is computed to make these contracts to be worth zero at inception. This fair price is known as the **forward/future price**. If the underlying asset is a stock that does not pay dividends, this forward price at time t is given by

$$F_t = S_t e^{r(T-t)},$$

where S_t denotes the asset price at time t, r is the interest rate, and $T - t$ is time to maturity. In the case of assets generating an income, this forward price has to be corrected to compensate these incomes.

At any time u before maturity, the forward/futures contract, itself, has a value, given (for the long position and when the underlying is a non-paying dividend stock) by

$$S_u - F e^{r(T-u)}.$$

Again, in the case of assets genereting an income, this price has to be corrected accordingly.

2.6 EXERCISES

1. Suppose that you enter into a 8-month forward contract on a non-dividend paying stock when the stock price is 90€ and the annual risk-free interest rate (with continuous compounding) is 5%. What is the forward price? **Solution: 93.05 €**

2. Suppose that you enter into a 6-month forward contract on a stock that will pay a dividend of 3€ in 4 months. If the the stock price is

80€ and the annual risk-free interest rate (with continuous compounding) is 1%. What is the forward price ? **Solution: 77.40 €**

3. Suppose that you enter into a 6-month forward contract on a currency that is worth 15€ today. If the the local annual risk-free interest rate (with continuous compounding) is 1% and the foreign interest rate is 2%, what is the forward price? **Solution: 14.93 €**

4. Under the scenario of Exercise 1, assume that after two months, the asset price is equal to 100. What is the value of the long forward after these two months? **Solution: 4.59 €**

5. In the scenario of Exercise 2, assume that after one month, the asset price is equal to 90. What is the value of the long forward after this month? **Solution: 10.28 €**

6. In the scenario of Exercise 3, assume that after three months, the asset price is equal to 90. What is the value of the long forward after these three months? **Solution: 0.037 €**

CHAPTER 3

Options

Options are contracts where the owner has the right (but not the obligation) to buy or to sell the underlying asset at some delivery price at some moment in the future. Obviously, this owner will exercise this right only if this leads to a positive income, and then the payoff of an option is never negative.

Even when they have similarities with forwards, options are more complex financial instruments. The main difference is that options are not 'fair' contracts as forward prices, but rights that can be bought or sold. Therefore, in this context there is no notion of 'fair' strike price, and options with different strike prices are traded and have a price.

3.1 CALL AND PUT OPTIONS

Call options give the right to buy, and **put** options give the right to sell, an underlying asset at some prefixed price K (the **strike price**) at some moment in the future. If the option can be exercised at any time before the maturity time T of the contract, we say that the option is **American**. If it can be exercised only at maturity, we say the option is **European**. American calls and puts on standard underlying assets (such as stocks, indexes, commodities, or currencies) are the simplest examples of options, and are called **vanilla** options. Options that are not vanillas are called **exotics**. The owner of the option is said to take a **long position**, while the seller of the option takes a **short position**.

Assume we have a long European call. If at maturity the asset price S_T is greater than K, we will exercise the option, and we will get a payoff equal to $S_T - K$. But, if $S_T < K$, we will not exercise it, and the final payoff will be equal to zero. Then, the payoff of a European long call is

given by
$$\max(S_T - K, 0).$$

In consequence, if we have a European short call, the final payoff is
$$-\max(S_T - K, 0).$$

Assume now we have a long European put option. If at maturity the asset price S_T is less than K, we will exercise the option, and we will get a payoff equal to $K - S_T$. But, if $S_T > K$, we will not exercise it, and the final payoff will be equal to zero. Then, the payoff of a European long put is given by
$$\max(K - S_T, 0).$$

In consequence, if we have a European short put, the final payoff is
$$-\max(K - S_T, 0).$$

The payoff functions of the long and the short positions of European calls and puts can be seen in Figure 3.1.

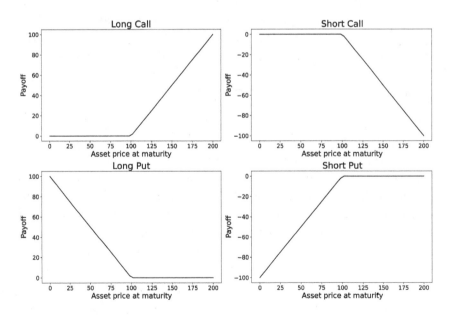

Figure 3.1 Payoffs of the long position and short positions of vanilla options with delivery price $F = 100$, in terms of the underlying asset price at maturity.

Notice that the profit of an option is not the same as the payoff. As the option has a price, the profit from a (long) option is the payoff minus the premium we paid for it.

3.2 THE INTRINSIC VALUE OF AN OPTION

The price of an option can be decomposed into two parts:

- The **intrinsic value**: this is the value of the option if it was exercised today. That is, the intrinsic value of a long European call at time t is $S_t - K$, while the intrinsic value of a long European put is $K - S_t$.

- The **time value** is the value of the option minus the intrinsic value.

Notice that, if t is very close to maturity, the asset price S_t is close to S_T, which implies that the payoff, the intrinsic value, and the option price are close quantities (and, in consequence, the time value is small). For longer maturities, its time value is higher and the value of the option differs from its intrinsic value.

If the intrinsic value of an option is positive, we say that the option is **in-the-money** (ITM). If this value is zero, we say the option is **out-of-the-money** (OTM). If $S_t = K$, the option is **at-the-money spot** (ATMS), while if $F_t = K$ the option is **at-the-money forward** (ATMF). In this book, **at-the-money** (ATM) will denote ATMF. Notice that, as OTM and ATMS have no intrinsic value, their price is given by their time value.

3.3 SOME PROPERTIES OF OPTION PRICES

Vanilla option prices are observed in the market for different maturities and strikes. Quantitative analysts take these values as inputs, and look for a model reproducing these prices, that is used to price exotics. We have to understand this model not as a model for asset price dynamics but an artefact that can explain observed vanilla prices and that allows us to determine the non-arbitrage price of exotics.

Even when vanilla prices are not computed but observed in the market, these prices have to satisfy some properties to avoid arbitrage. Let us see how, from the payoff function, we can deduce some properties of option prices and what the impact of each market parameter on them.

In order to fix ideas, let us assume that $S_t = 100$ and let us consider the fictitious market option prices for different maturities and strikes in Tables 3.1 and 3.2.

Table 3.1 Fictitious call prices for different maturities (in years) and strikes (in euros).

Time to maturity/strike	80	90	100	110	120
0.1	20.03	10.60	3.78	0.82	0.11
0.5	21.43	13.99	8.45	4.75	2.50
1	23.53	17.01	11.92	8.14	5.44

Table 3.2 Fictitious put prices for different maturities (in years) and strikes (in euros).

Time to maturity/strike	80	90	100	110	120
0.1	0.03	0.60	3.78	10.82	20.10
0.5	1.43	3.99	8.45	14.75	22.50
1	3.53	7.01	11.92	18.14	25.44

3.3.1 The Price of an Option vs the Price of an Asset

Let us take a first look at the option prices in Table 3.1. The first we observe is that call prices are really cheaper than the underlying asset. Why? The answer follows directly from the observation of the corresponding payoff functions: for the European call with strike K, this payoff is
$$\max(S_T - K, 0),$$
while the payoff of a long position in assets is S_T. As the payoff of the call is less, today's price has to be less, according to the data in Table 3.1. This property has relevant implications, as we will see in this book. As investing in call options is cheaper than investing in assets, options become an interesting alternative when traders look for strategies with a minimum investment.

In a similar way, a put option has a payoff equal to

$$\max(K - S_T, 0) < K,$$

and then, its price is always smaller than the strike price, as we can observe in Table 3.2.

3.3.2 The Role of the strike price

From Tables 3.1 and 3.2, we can observe that call prices decrease with the strike, while put prices increase with them. Let us see why. Consider two European calls with strike prices $K_1 > K_2$. Notice that

$$\max(S_T - K_1, 0) < \max(S_T - K_2, 0).$$

Then, the payoff of the option with the lower strike is higher, which implies that this option has to be more expensive. In a similar way, if we consider two European puts with strike prices $K_1 > K_2$, the payoffs satisfy that

$$\max(K_1 - S_T, 0) > \max(K_2 - S_T, 0).$$

Now, the payoff is higher for the lower strike, and then this option has to be the most expensive. That is, European call (put) prices are decreasing (increasing) functions of the strike price.

3.3.3 The Role of the Price of the Underlying Asset

Consider a European call option with payoff

$$\max(S_T - K, 0).$$

It is natural to assume that the higher the asset price S_t, the higher the expected value for S_T, and then the higher the expected value for $\max(S_T - K, 0)$. Then, European call prices have to be increasing as functions of the asset price S_t. In the case of put options, the scenario is just the opposite, and European put prices are decreasing as functions of the underlying asset price S_t. In conclusion, European call (put) prices are increasing (decreasing) functions of the asset price. We can observe this phenomenon in Figure 3.2.

30 ■ Options

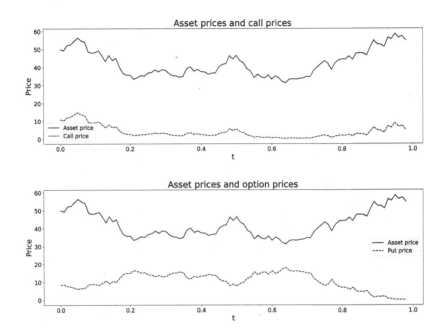

Figure 3.2 Simulation of the evolution of asset prices and ATM call prices (top), and asset prices and ATM put prices (bottom), with $T = 1$ year.

3.3.4 The Role of Interest Rates

Call options are more expensive when interest rates are high, and put prices are more expensive when interest rates are low, as we can observe in the dummy market prices in Table 3.3.

Table 3.3 Dummy call prices for different interest rates (cc).

Option type / Interest rate	0	0.02	0.04	0.06	0.08	0.1
Call	11.92	12.82	13.75	14.71	15.71	16.73
Put	11.92	10.84	9.83	8.89	8.02	7.21

Why? In general, it is not interesting to buy when interest rates are high. If we have to borrow money, this will lead to huge interest payments. If we have the money, investing it in an interest-bearing account will give us a big interest income. In this scenario, buying an asset is expensive, and traders look for cheaper financial instruments.

The price of call options is less than the price of the underlying asset (as we discussed in Section 3.3.1) and they can lead to a big profit. If interest rates are high, they become interesting instruments for traders. Then, as the demand of options increase, their price increase, as we can observe in Table 3.3.

In the case of a put option, the scenario is the opposite: by shorting the asset, we receive an income that we can invest in a saving account, leading to a benefit due to interest rates. Then, the price of a put is a decreasing function of interest rates.

3.3.5 The Role of Volatility

Option prices (both calls and puts) have a non-negative payoff. In this context, higher volatilities increase the probability of getting a big payoff, which increases the price of the option, both for calls and puts. That is, option prices are an increasing function of volatility.

Notice that the volatility affects only the time value of an option, and not its intrinsic value. Then, the impact of volatility will be more relevant in options with a higher time value, such as long-term options. The sensitivity of prices with respect to volatility can be observed in the following fictitious option prices (here we assume $S_t = K = 100$, $r = 0.0$).

Table 3.4 Fictitious call prices for different interest rates (cc).

Time to maturity/volatility	0.01	0.02	0.03
0.1	0.12	0.25	0.38
1	0.40	0.80	1.20

3.3.6 The Role of Time to Maturity

Time to maturity has a similar effect as high volatilities. Long maturities increase the probabilities of getting a huge payoff, and then options with long maturities (both calls and puts) tend to be more expensive (see Figures 3.1 and 3.2). This phenomena can be observed in Figure 3.4.

Nevertheless, this general rule is not always true. Consider for example the case of a deep ITM put option, where the asset price is close to zero and very small in comparison with the strike price. In this scenario, the payoff $\max(K - S_T, 0)$ will be closer to K, and then the option price will be closer to the present value of K, which (if interest rates are positive) is a decreasing function of time to maturity.

Very short maturities imply that the price of the option is close to its intrinsic value, while for long-term options, the time value is higher and the option price differs from this intrinsic value. This can be observed in the data in Tables 3.5 and 3.6, where we have added the intrinsic value to the dummy data in Tables 3.1 and 3.2.

Table 3.5 Intrinsic values and fictitious call prices for different maturities (in years) and strikes (in euros).

Time to maturity/strike	80	90	100	110	120
Intrinsic value	20	10	0	0	0
0.1	20.03	10.60	3.78	0.82	0.11
0.5	21.43	13.99	8.45	4.75	2.50
1	23.53	17.01	11.92	8.14	5.44

Table 3.6 Intrinsic values and fictitious put prices for different maturities (in years) and strikes (in euros).

Time to maturity/strike	80	90	100	110	120
Intrinsic value	0	0	0	10	20
0.1	0.03	0.60	3.78	10.82	20.10
0.5	1.43	3.99	8.45	14.75	22.50
1	3.53	7.01	11.92	18.14	25.44

3.3.7 The Put-Call Parity

Call and put prices are not independent. Notice that

$$\max(S_T - K, 0) - \max(K - S_T, 0) = S_T - K,$$

as we can observe in Figure 3.3.

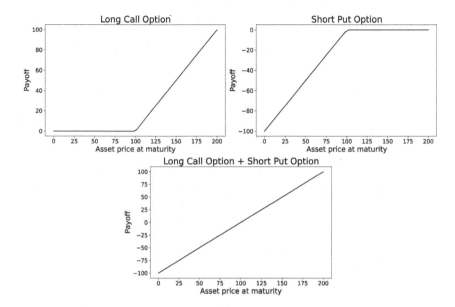

Figure 3.3 Equivalence between the payoff of a portfolio with a long call and a short put with $K = 100$ and a long forward contract.

Then, a portfolio that consists of a long call and a short put (with the same strike K and maturity time T) has the same payoff as a long forward with delivery price K. As a consequence, their price has to be the same today.

The value of this portfolio is $c - p$, where c denotes the price of the call and p is the price of the put. On the other hand, if the underlying asset is a non-paying dividend stock, the value of the long forward is given by today's price of the asset minus the present value of the strike, $S_t - e^{-r(T-t)}K$. Both the portfolio and the long forward have to have the same price:

$$c - p = S_t - e^{-r(T-t)}K,$$

a relation that is called the **put-call relationship**. Notice that, in absence of arbitrage, the price of a call is determined by the price of a put and vice versa.

3.4 SPECULATION WITH OPTIONS

Option prices are lower than asset prices, and then they become interesting in speculation. To buy them may not cost very much, especially in the case of OTM that are close to maturity. If the option is worthless at maturity, the loss is small, but if there is a big movement in the asset price, the option can expire ITM, leading to a large profit. Let us see this in the following example

Example 3.1

Assume the same data as in Table 3.1, and consider a call with strike $K = 110$ and time to maturity $T = 0.5$ years, that is worth 4.75 euros. Let us compare the results of two strategies (to buy an asset/to buy an option) in different scenarios:

1. *First scenario: at maturity, the asset price is equal to 120.*

    - To buy 1 asset: The final profit of this strategy is
    $$\frac{(120 - 100)100\%}{100} = 20\%$$
    of the initial investment.

    - To buy a European call. The final profit of this strategy is
    $$\frac{(120 - 110 - 4.75)100\%}{4.75} = 110.52\%$$
    of the initial investment.

2. *Second scenario: at maturity, the asset price is equal to 90.*

    - To buy 1 asset: The final profit of this strategy is
    $$\frac{(100 - 90)100\%}{100} = -10\%$$
    of the initial investment.

    - To buy a European call. The final profit of this strategy is
    $$\frac{-4.75 \times 100\%}{4.75} = -100\%$$
    of the initial investment.

That is, the strategy based on OTM options can lead to a higher profit (in % of the original investment), but it can also lead to a 100% loss, becoming then a more risky strategy than being long in a stock. Notice that the risk is higher in short positions because the loss is not limited to the premium of the option.

3.5 SOME CLASSICAL STRATEGIES

The combination of calls and puts allow us to construct strategies with different payoffs. As examples, let us introduce two of them: the bull and the bear spreads.

3.5.0.1 Bull Spread

Imagine for example, that we buy a call with strike 100 and we write a call with strike 150. Then, the final payoff of this strategy is given by

$$\max(S_T - 100, 0) - \max(S_T - 150, 0).$$

Notice that this quantity is zero if $S_T < 100$, it is equal to $S_T - 100$ if $S_T \in (100, 150)$, and it is equal to 50 if $S_T > 150$. This strategy is called a bull spread, as we can see in Figure 3.4.

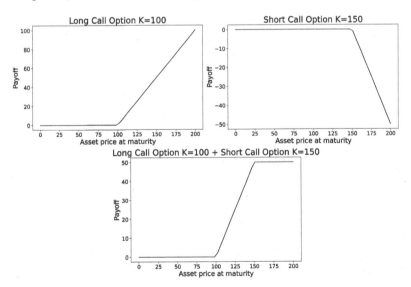

Figure 3.4 A bull spread is composed of a long call with $K_1 = 100$ and a short call with $K_2 = 150$.

In general, a bull strategy can be constructed with a long call with strike K_1 and a short call with strike K_2, where $K_2 > K_1$. This strategy is interesting in the context of a bull (rising) market. Notice that the price of a bull spread is less than the price of a long call, since its payoff is less.

Remark 3.5.1 *A bull spread can also be constructed by buying a put option with K_1 and selling a put with K_2, where $K_2 > K_1$.*

3.5.0.2 Bear Spread

Imagine for example, that we sell a put with strike 100 and we buy a put with strike 150. Then, the final payoff of this strategy is given by

$$\max(150 - S_T, 0) - \max(100 - S_T, 0).$$

It is easy to see that this quantity is 50 if $S_T < 100$, it is equal to $150 - S_T$ if $S_T \in (100, 150)$, and it is equal to 0 if $S_T > 150$. This strategy is called a **bear spread**. It benefits from a bear (falling) market (see Figure 3.5).

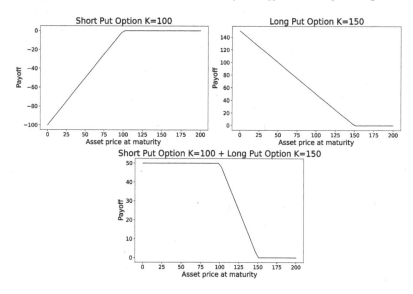

Figure 3.5 A Bear Spread is composed of a short put with $K_1 = 100$ and a long put with $K_2 = 150$.

In general, a bear spread can be constructed with a short put with strike K_1 and a long put with strike K_2, with $K_1 > K_2$. Notice that the price of a bear spread is less than the price of a long put. Moreover, a bear spread can also be constructed with a short call with strike K_1 and a long call with strike K_2, with $K_1 < K_2$.

3.6 DRAW YOUR STRATEGY WITH PYTHON

In this chapter, we have learn about how options are with a special focus on call and put options. We have also seen that we can create different payoffs combining the options. For example, in the put-call parity, we observe that buying a call option and selling a put option ATMF give us a forward. Let's see how to translate this to Python. Thus, we can draw any strategy that we imagine. We will start for drawing the long and short forward payoff. The first step is to define the long and short forward. In this case, we impose the forward is 100.

```
#Packages needed

#Numerical package
import numpy as np

#Graphical Package
import matplotlib.pyplot as plt

#We define a Long Forward Payoff
#with delivery price 100
def payofflongforward(S):
    return S-100

#We define a Short Forward Payoff
#with delivery price 100
def payoffshortforward(S):
    return 100-S
```

Now, we can proceed to make the first plot based on two subplots: one with the long forward and another with the short of one.

```
#We create a equispaced vector with
#values from 0 to 200
S=np.linspace(0,200)

#We create a figure to insert the plots
plt.figure(figsize =(15 ,7))

#In the first plot
plt.subplot(1,2,1)

#We write the title
```

```python
plt.title("Long Forward", fontsize=22)

#We write a x-axis label
plt.xlabel("Asset price at maturity", fontsize=18)

#We write a y-axis label
plt.ylabel("Payoff", fontsize=18)

#We do the plot of the Long Forward
plt.plot(S,payofflongforward(S), color="black")

plt.xticks(fontsize=14)
plt.yticks(fontsize=14)

#The second plot
plt.subplot(1,2,2)

#We write the title
plt.title("Short Forward", fontsize=22)

#We write a x-axis label
plt.xlabel("Asset price at maturity", fontsize=18)

#We write a y-axis label
plt.ylabel("Payoff", fontsize=18)

#We do the plot of the Short Forward
plt.plot(S,payoffshortforward(S), color="black")

plt.xticks(fontsize=14)
plt.yticks(fontsize=14)

plt.show()
```

The result of this code is Figure 2.1. Now, we want to make things more complicated. We want to calculate the four combinations of long and short positions for a call and a put option. Again, we start by defining the payoff functions.

```python
#Packages needed

#Numerical package
import numpy as np

#Graphical Package
import matplotlib.pyplot as plt

#We define a Long Call
#with Strike 100
```

```
def LongCall(S):
    return np.maximum(S-100,0)

#We define a Long Put
#with Strike 100
def LongPut(S):
    return np.maximum(100-S,0)
```

Now, we can proceed to create Figure 3.1.

```
#We create a equispaced vector with
#values from 0 to 200
S=np.linspace(0,200)

#We create a figure to insert the plots
plt.figure(figsize =(15 ,10))

#In the first plot
plt.subplot(2,2,1)

#We write the title
plt.title("Long Call", fontsize=22)

#We write a x-axis label
plt.xlabel("Asset price at maturity", fontsize=18)

#We write a y-axis label
plt.ylabel("Payoff", fontsize=18)

#We do the plot of the Long Forward
plt.plot(S,LongCall(S), color="black")

#size of the axis numbers
plt.xticks(fontsize=14)
plt.yticks(fontsize=14)

#The second plot
plt.subplot(2,2,2)

#We write the title
plt.title("Short Call", fontsize=22)

#We write a x-axis label
plt.xlabel("Asset price at maturity", fontsize=18)

#We write a y-axis label
plt.ylabel("Payoff", fontsize=18)

#We do the plot of the Short Forward
```

```python
plt.plot(S,-LongCall(S), color="black")

#size of the axis numbers
plt.xticks(fontsize=14)
plt.yticks(fontsize=14)

#In the first plot
plt.subplot(2,2,3)

#We write the title
plt.title("Long Put", fontsize=22)

#We write a x-axis label
plt.xlabel("Asset price at maturity", fontsize=18)

#We write a y-axis label
plt.ylabel("Payoff", fontsize=18)

#We do the plot of the Long Forward
plt.plot(S,LongPut(S), color="black")

#size of the axis numbers
plt.xticks(fontsize=14)
plt.yticks(fontsize=14)

#The second plot
plt.subplot(2,2,4)

#We write the title
plt.title("Short Put", fontsize=22)

#We write a x-axis label
plt.xlabel("Asset price at maturity", fontsize=18)

#We write a y-axis label
plt.ylabel("Payoff", fontsize=18)

#We do the plot of the Short Forward
plt.plot(S,-LongPut(S), color="black")

#size of the axis numbers
plt.xticks(fontsize=14)
plt.yticks(fontsize=14)

#We make enough space between the subplots
plt.tight_layout()

plt.show()
```

If we want to create strategies for buying and selling options with different strikes, we will need to define as many options as there are strikes. That is not very functional. Instead, we can re-define the option's payoff to incorporate the strike as a variable.

```python
#Packages needed

#Numerical package
import numpy as np

#Graphical Package
import matplotlib.pyplot as plt

#We define a Long Call for any Strike
def LongCallv2(S,K):
    return np.maximum(S-K,0)

#We define a Long Put for any Strike
def LongPutv2(S,K):
    return np.maximum(K-S,0)
```

Now, we are going to replicate the Figure 3.4.

```python
#We create a equispaced vector with
#values from 0 to 200
S=np.linspace(0,200)

#We create a figure to insert the plots
plt.figure(figsize =(15 ,10) )

#In the first plot
plt.subplot(2,2,1)

#We write the title
plt.title("Long Call Option with K=100", fontsize=22)

#We write a x-axis label
plt.xlabel("Asset price at maturity", fontsize=18)

#We write a y-axis label
plt.ylabel("Payoff", fontsize=18)

#We make the Long Call plot
plt.plot(S,LongCallv2(S,100), color="black")

#In the second plot
plt.subplot(2,2,2)

#We write the title
```

```
plt.title("Short Call Option with Strike K=150", fontsize=22)

#We write a x-axis label
plt.xlabel("Asset price at maturity", fontsize=18)

#We write a y-axis label
plt.ylabel("Payoff", fontsize=18)

#We make the Short Call plot
plt.plot(S,-LongCallv2(S,150), color="black")

#In the third plot
plt.subplot(2,2,3)

#We write the title
plt.title("Long Call Option with K=100 \
        + Short Call Option with Strike K=150", fontsize=22)

#We write a x-axis label
plt.xlabel("Asset price at maturity", fontsize=18)

#We write a y-axis label
plt.ylabel("Payoff", fontsize=18)

#We make the Strategy plot
plt.plot(S,LongCallv2(S,100)-LongCallv2(S,150), color="black")

#We make enough space between the subplots
plt.tight_layout()

plt.show()
```

3.7 CHAPTER'S DIGEST

Call (or put) options are financial instruments that allow their holders to sell (buy) something at some future moment and at some prefixed price. Options that can be exercised only at maturity are called European, while if they can be exercised at any time before maturity, they are called American. European and American options are the simplest examples of options, and we refer to them as vanillas. Options that are not vanillas are called exotics, and they include barrier, lookback and barrier options.

Options are instruments that can be bought and sold, and then they have a price. Vanilla option prices are observed in the market and they strongly depend on several parameters as asset price, interest rate of volatility, as well as on strike and time to maturity. Moreover, option prices have to satisfy the call-put parity, a relationship between the price of a European call and the price of the corresponding European put.

We can use options to construct more complex strategies such as bull or bear spreads.

3.8 EXERCISES

1. Assume that $S_0 = 100$ and that the price of a European call with $K = 100$ and $T = 1$ is equal to 14.2312. If the price of the European put with the same strike and maturity is 9.3541, what is the interest rate? **Solution: 0.05 (cc)**.

2. Assume that $S_0 = 100$, and that the price of a European call with $K = 100$ and $T = 1$ is equal to 12.8215. If the interest rate is equal to 2% (cc), what is the price of a European put with the same strike and time to maturity? **Solution: 10.8414**.

3. A **butterfly** is a strategy composed of a long call with strike K_1, two short calls with strike K_2, and a long call with strike K_3, with $K_2 = \frac{K_1 + K_3}{2}$, and where all the maturities are the same. Plot the payoff of this strategy.

4. A **long straddle** is a strategy composed of a long call and a long put with the same strike. Plot the payoff of this strategy.

5. A **covered call** is composed of a short call and an asset. Plot the payoff of this strategy.

CHAPTER 4

Exotic Options

We have seen in Chapter 3 that European or American calls and puts on standard underlying assets are called vanillas. More complex options are named exotics, and they are usually traded OTC. Let us see in this chapter the main types of exotic options.

4.1 BINARY OPTIONS

The holder of a **binary option**, also named *all-or-nothing* option, receives a monetary amount if some event happens at maturity. If not, the payoff is zero. They can be:

- **Cash-or-nothing options:** The holder of the option receives a fixed quantity if the option expires ITM, while nothing if it expires OTM.

- **Asset-or-nothing options:** The holder of the option receives the final value of the underlying asset if the option expires ITM, while nothing if it expires OTM.

The plots of the payoff of an all-or-nothing option that pays 30€ if $S_T > 100$ and for a cash-or-nothing option that pays S_T if $S_T > 100$ can be seen in Figure 4.1.

46 ■ Exotic Options

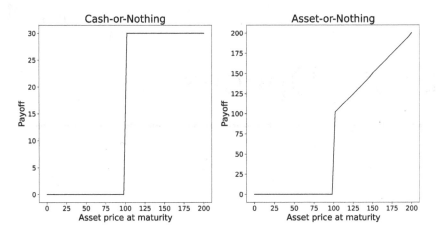

Figure 4.1 Payoffs of a all-or-nothing option that pays 30 euros if $S_T > 100$ and for a cash-or-nothing option that pays S_T if $S_T > 100$.

4.2 FORWARD START OPTIONS

Roughly speaking, a **forward start option** is an European option that starts in the future, at some moment T_1 between inception and maturity. There are two types of forward start options:

- Options whose strike is fixed at T_1. Usually, this strike is a function of S_{T_1}. For example, their payoff is written as

$$\max(S_T - cS_{T_1}, 0)$$

  where c is a constant that we fix in the contract. Several times, c is assumed to be equal to 1.

- Options with a payoff that depends on the ratio between S_T and S_{T_1}. That is, with a payoff of the form

$$\max\left(\frac{S_T}{S_{T_1}} - K, 0\right).$$

A sequence of forward start options (that is, one option starts with the previous one expires) is called a **cliquet**. Notice that, when we arrive at T_1, S_{T_1} is a constant and then the above two types of options 'start' as classical vanillas. Moreover, if $c = K = 1$, these options are (at T_1) At-The-Money spot options.

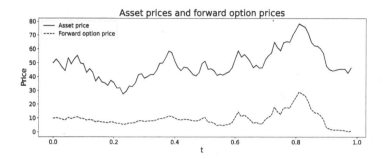

Figure 4.2 Simulation of the evolution of asset prices and the price of a forward start option with payoff $\max(S_T - S_{T_1}, 0)$, where $T = 1$, and $T_1 = 0.5$ (in years).

The behavior of the price of a forward start option can be seen in Figure 4.2. Notice the change of regime at time $t = T_1 = 0.5$, when the call option starts.

4.2.1 Compound Options

Compound options are options on options (that is, the underlying asset is an option). Notice that the maturities and the strikes of the underlying option and the compound option do not have to be the same. Consider, for example, a call (with maturity T_1 and strike K_1) on a call with maturity $T_2 > T_1$ and strike K_2. Then, the payoff of the compound option is given by

$$\max(V_{T_1} - K_1, 0),$$

where V_{T_1} is the value, at time T_1, of a call with maturity T_2 and strike K_2.

As options are cheaper than the underlying asset, a compound option is cheaper than a classical vanilla option, and then compound options tend to be very cheap, as we can see in Figure 4.3.

4.3 PATH-DEPENDENT OPTIONS

In path-dependent options, the final payoff depends not only on the final value of the underlying asset, but on the price of the asset during the whole life of the option. In this section, we study some examples of path-dependent options such as barrier, lookback or Asian.

48 ■ Exotic Options

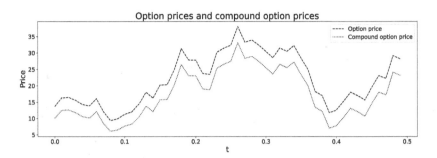

Figure 4.3 Simulation of the evolution of call prices (with $K_2 = 50$ euros and $T = 1$ year), and a call on this call, with $K_1 = 5$ and $T = 0.5$.

4.3.1 Barrier Options

In a **barrier option**, the payoff depends on the price of the underlying reaching or not a certain level (barrier) before maturity. There are different types of barrier options.

- **Knock-in-options**, where the option can be exercised if the asset price reached the barrier.
  - **Down-and-in**: At inception, the asset price is higher than the barrier, and this price has to go down to reach the barrier.
  - **Up-and-in**: At inception, the asset price is lower than the barrier, and this price has to go up to reach the barrier.
- **Knock-out-options**, where the option can be exercised if the asset prices never reached the barrier before maturity.
  - **Down-and-out**: At inception, the asset price is higher than the barrier.
  - **Up-and-out**: At inception, the asset price is lower than the barrier.

Barrier options are cheaper than the corresponding vanilla options. If the market has low volatility, one can expect that the asset price will not reach a 'far enough' barrier, and a knock-out option can be interesting in this scenario. If the market volatility is high, one can expect the volatility to reach some 'close enough' barrier, and then a knock-in-option can be convenient. We can see an example of an asset with high volatility. In this case, the probability of crossing the barrier is high, and a down-in-option will be convenient, while a down-out option not, as we can see in Figures 4.4 and 4.5.

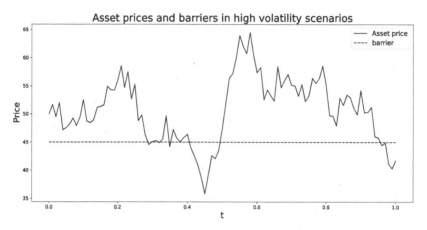

Figure 4.4 Simulation of asset prices with high volatility. Notice that the probability of crossing the barrier is high. In this context, a knock-in option can be convenient.

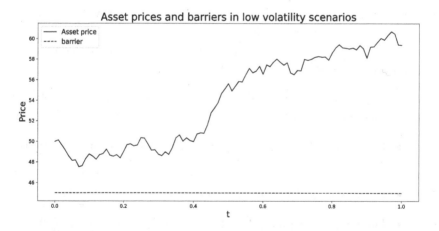

Figure 4.5 Simulation of asset prices with low volatility. Notice that the probability of crossing the barrier is low. In this context, a knock-out option can be convenient.

4.3.2 Lookback Options

In **lookback options** the payoff depends on the optimal (maximum or minimum) value of the asset price before maturity. For example:

- The payoff of a **floating lookback call** is given by $S_T - S_{\min}$, where $S_{\min}$ denotes the minimum of S during the life of the option.

- A **floating lookback put** pays $S_{\max} - S_T$, where $S_{\max}$ denotes the maximum of S from inception to maturity.

- A **fixed lookback call** pays $\max(S_{\max} - K, 0)$, for some fixed strike K.

- The payoff of a **fixed lookback put** is $\max(K - S_{\min}, 0)$, where K is some fixed strike.

The payoff of a floating lookback option is always positive. Then these options are always exercised, and then they are more expensive than classic vanillas. On the other hand, the payoff of a fixed lookback option is higher than the corresponding vanilla, and then they are also more expensive. Let us see a possible outcome in the following example.

Example 4.1

Consider the simulation of asset prices in Figure 4.6, where we can also see the evolution of the maximum and minimum prices. At maturity, the asset price is equal to 63.4, the maximum is equal to 124.4 and the minimum is 48.2. Then

- The payoff of a vanilla call with strike equal to 100 is $\max(63.5 - 100, 0) = 0$.

- The payoff of a fixed lookback call with strike equal to 100 is $\max(124.4 - 100, 0) = 24.4$.

- The payoff of a floating lookback call is $63.5 - 48.2 = 15.3$.

- The payoff of a put call with strike equal to 100 is $\max(100 - 63.5, 0) = 16.5$.

- The payoff of a fixed lookback put with strike equal to 100 is $\max(100 - 48.2, 0) = 51.8$.

- The final payoff of a floating lookback put is $124.4 - 63.5 = 60.9$.

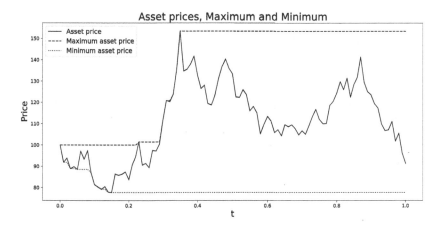

Figure 4.6 Simulation of asset prices and their maximum and minimum.

4.3.3 Asian Options

In Asian options, the payoff depends on the arithmetic mean of asset prices $\bar{S}$ during the live of the option. Some examples are:

- **Fixed strike call**, with payoff $\max(\bar{S} - K, 0)$,
- **Fixed strike put**, with payoff $\max(K - \bar{S}, 0)$,
- **Floating strike call**, with payoff $\max(S_T - \bar{S}, 0)$,
- **Floating strike put**, with payoff $\max(\bar{S} - S_T, 0)$.

Let us see a possible outcome in the following example.

> **Example 4.2**
>
> Consider the simulation of asset prices in Figure 4.7, where we can also see the evolution of the average. At maturity, the asset price is equal to 105.9 and the average is 120.4. Then
>
> - The payoff of a vanilla call with $K = 100$ is $\max(105.9 - 100, 0) = 5.9$.
> - The payoff of a vanilla put with $K = 100$ is $\max(100 - 105.9, 0) = 0$.
> - The payoff of a fxed strike call with $K = 100$ is $\max(120.4 - 100, 0) = 20.4$.

- The payoff of a fixed strike put with $K = 100$ $\max(100 - 120.4, 0) = 0$.

- The payoff of a floating strike call is $\max(105.9 - 120.4, 0) = 0$.

- The payoff of a floating strike put is $\max(120.4 - 105.9, 0) = 14.5$.

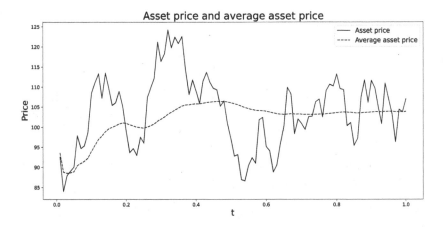

Figure 4.7 Simulation of asset prices and their mean.

As the variance of the average price is less than the variance of the asset price at maturity, Asian options are cheaper than the corresponding vanillas.

4.4 SPREAD AND BASKET OPTIONS

In **spread options**, the payoff depends on the difference ('spread') between two asset prices S_T^1 and S_T^2 at maturity. That is, the payoff of a spread option is of the type

$$\max(S_T^1 - S_T^2 - K, 0).$$

A spread option with $K = 0$ is called a **exchange option**, and its payoff reduces to

$$\max(S_T^1 - S_T^2, 0).$$

In **basket options**, this payoff depends of a set ('basket') of options.

Spread options are common in commodity markets, where spreads appear as the difference in prices between the assets that are inputs and outputs in a production process. For example, a **spark spread** is the difference between the price of electricity and the cost of production from natural gas.

4.5 BERMUDA OPTIONS

European options can be exercised only at maturity, while American options are exercisable during the whole life of the option. **Bermudan options** are between European and American options in the sense that they have a set of exercise dates, typically some fixed days.

Obviously, the Bermuda option is more expensive than the corresponding European option, and cheaper than the corresponding American option.

4.6 CHAPTER'S DIGEST

Non-vanilla options are called exotics. There are several types of exotic options that allow the traders to create taylor-made strategies in different scenarios. Classical examples of exotics include binary, forward start, compound, path-dependent (as barrier, lookback or Asian), spread, and Bermuda.

4.7 EXERCISES

- Explain how can we approximate a binary call by a sequence of bull spreads.

- Explain which options could be interesting in a high-volatility scenario.

- Explain which options could be interesting in a low-volatility scenario.

- Give some examples, in commodity markets, where a spread option may be convenient.

CHAPTER 5

The Binomial Model

In the previous chapters, we have introduced what an option is, from the most classic and simple ones, such as vanilla options, to a wide variety of modifications, such as exotic options. But one of the most natural and crucial questions remains unanswered: what is the option price?

In the case of futures (or forwards), we have seen that an arbitrage-free price can be constructed by combining the asset and some cash. This process is called a replicating portfolio. This portfolio behaves exactly like the future (or forward). Unfortunately, it is impossible to find a similar formula when the payoff function is a non-linear function like $(S_T - K)_+$. To find a price in this situation, we need to assume the random behavior of the underlying.

Several ways to estimate the future behavior of an asset can come to our mind, but we are going to start from the simplest: the binomial model. In this model, today's price can change to two different prices at a future time. Moreover, one price will increase, and the other will decrease. Under this model, we can replicate the option behavior over a short period. This toy model is helpful to understand how to price and hedge options, but in terms of asset price dynamics, it is a poor model. This model was introduced in 1979 in (Cox et al., 1979).

5.1 THE SINGLE-PERIOD BINOMIAL MODEL

In the market, an asset is trading today with a price of S_t. We are interested in buying a European option that will expire in a week. It is impossible to know the asset price at the option expiration. Assume that the asset price can take two possible values in a week. One of the two possible values will be a price increase above the risk-free rate with a probability of p, while the other will be a price decrease below the risk-free rate with a probability of $1-p$. As there are two possible outcomes, the sum of the probabilities of the two events must be 1.

For example, the asset price today is 10€. In the future, the price of the asset may reach 12 or 8€ (see Figure 5.1).

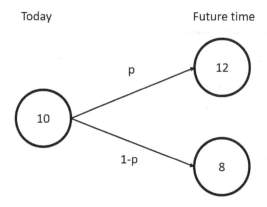

Figure 5.1 Example Binomial Model.

We can generalize the model, by assuming that the stock can increase $S_t u$ and decrease $S_t d$. Note that u is for the upward trend and d for the downward trend (see Figure 5.2).

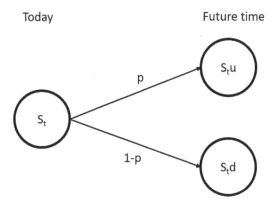

Figure 5.2 General Binomial Model.

If we decide to buy a call option with strike 11, we will have the following situation.

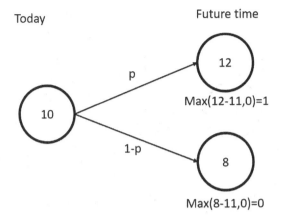

Figure 5.3 Call option Payoff in the Binomial Model.

We can generalize in the following way (see Figure 5.4).

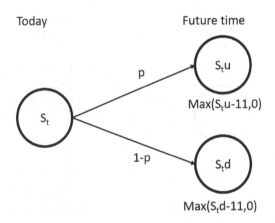

Figure 5.4 General Call option Payoff in the Binomial Model.

Instead, if we decide to buy a put option with strike 11, we will have the following situation (see Figure 5.5).

58 ■ The Binomial Model

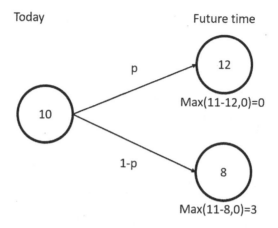

Figure 5.5 Put option Payoff in the Binomial Model.

We can generalize in the following way (see Figure 5.6).

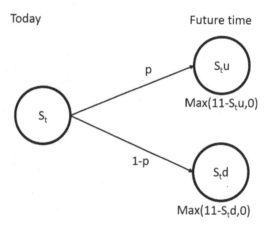

Figure 5.6 General Put option Payoff in the Binomial Model.

We see that once we find the future asset values, we can apply the payoff function to find the option value in each situation. It seems that the necessary elements to calculate the option price are the probability p, the increment value u, the decrement value d, and how to discount the future option value to the present time.

5.1.1 Relationship between European Options and Their Underlying in the Binomial Model

In Chapter 3.3.3, we have seen a relationship between the movement of an option and its underlying. For example, when the underlying increases, so the option does. At that time, we couldn't quantify that relationship, but now with our toy model, we can.

Imagine we want to trade a European call option with a strike of 11€. The underlying price is 10€. In the future, the underlying may reach 12 or 8€. As we have seen in Figure 5.3, the option price may be 1 or 0€. Therefore, there are two possibilities:

- The asset price is 12€, then the call option price is 1€.
- The asset price is 8€, then the call option price is 0€.

Note that we can draw a line between these two points (see Figure 5.7).

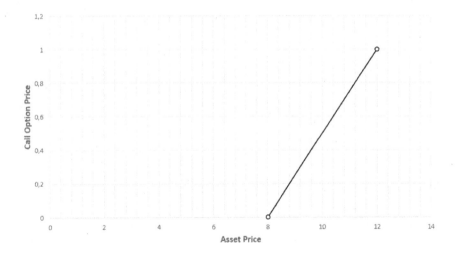

Figure 5.7 Relationship between the asset price and the option price.

This line tells us how many units of the option price change when the asset price changes. This relationship is defined by the line slope. We will refer to this slope with the Greek name Δ, and we can calculate it as:

$$\Delta \doteq \frac{\text{Total change on the option price}}{\text{Total change on the asset price}} = \frac{1-0}{12-8} = \frac{1}{4}. \quad (5.1)$$

In a general framework, we can write this formula for any European call option:

$$\begin{aligned}\Delta &= \frac{\text{Total change on the option price}}{\text{Total change on the asset price}} \\ &= \frac{\max(S_t u - K, 0) - \max(S_t d - K, 0)}{S_t u - S_t d} \\ &= \frac{\max(S_t u - K, 0) - \max(S_t d - K, 0)}{S_t (u - d)}. \end{aligned} \quad (5.2)$$

The call option's Δ is always bigger or equal to zero. This is because by definition $u > d$ and therefore $\max(S_t u - K, 0) - \max(S_t d - K, 0) \geq 0$.

In a similar way, we can write an analogue formula for any European put option:

$$\begin{aligned}\Delta &= \frac{\text{Total change on the option price}}{\text{Total change on the asset price}} \\ &= \frac{\max(K - S_t u, 0) - \max(K - S_t d, 0)}{S_t u - S_t d} \\ &= \frac{\max(K - S_t u, 0) - \max(K - S_t d, 0)}{S_t (u - d)}. \end{aligned} \quad (5.3)$$

Using the arguments as before, the put option's Δ is always less or equal to zero.

5.1.2 Replication Portfolio for European Options

We have seen that in the Binomial model, the asset evolution depends on some parameters. A natural way of finding the option price is to estimate them. It seems impossible to calculate the price as we did before with futures (or forwards). In the end, the possible option values are random in our model. But is there a way to remove the randomness?

We have seen that there is a relationship between the option price and its underlying. We can take advantage of this relationship to remove the randomness. To do that, we create a portfolio that combines the option and its underlying. The portfolio has the following trades:

- A European option.
- - Δ times the underlying asset.

The Single-Period Binomial Model ■ 61

Imagine we want to trade a European call option with a strike of 11€, maturity time 5 days. The underlying price is 10€. In the future, the underlying may reach 12 or 8€. Assume the risk-free interest rate is 5% flat. In this example, the call option Δ is 0.25 as seen in (5.1).

Now, if we calculate the portfolio value, we have two possibilities (see Figure 5.8):

- If the asset price is 12€, the option price is 1€ and the portfolio value is the option price - Δ times the asset price, i.e., $\max(12-11, 0) - 0.25 \cdot 12 = 1 - 3 = -2$.

- If the asset price is 8€, the option price is 0€ and the portfolio value is the option price - Δ times the asset price, i.e., $\max(8-11, 0) - 0.25 \cdot 8 = 0 - 2 = -2$.

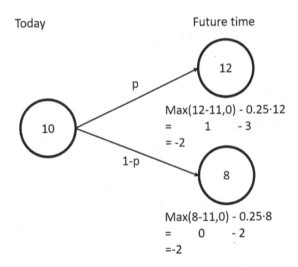

Figure 5.8 Replication of Call option.

Surprisingly, the portfolio value is the same in both scenarios. We have frozen the market movement. But in all cases, we receive a fixed amount of money. In this case, the amount is -2. Therefore, we can discount it with the risk-free interest rate to the initial time. Then, by no-arbitrage conditions, at the initial time, our portfolio value must be $-2 \cdot \exp\left(-5\% \cdot \frac{5}{365}\right)$. That means, we have at the initial time:

$$\text{Option price} - 0.25 \cdot 10 = -2 \cdot \exp\left(-5\% \cdot \frac{5}{365}\right). \quad (5.4)$$

It is an equation with one unknown. Using linear algebra, we find the option value.

$$\text{Option price} = \frac{5}{2} - 2 \cdot \exp\left(-5\% \cdot \frac{5}{365}\right). \tag{5.5}$$

In a one-step binomial model, we can rewrite the above equation as the combination of different assets:

$$\text{Option price} = \Delta \cdot S_t - \text{Cash}. \tag{5.6}$$

Thus, during a period, the correct combination of an option and its underlying behaves like cash. Or in other words, by the associative property, the proper combination of cash and an option's underlying behaves like the option. This is a model property. When options are replicable in a model, we say the market is complete.

Another important thing is that we have not needed any probability to obtain the option price. This is possible because we have been able to freeze market conditions. To do that, we are only interested in the width of the option and its underlying. The process of removing randomness from a portfolio is known as hedging.

We can restate the problem from a mathematical perspective. In the case of a call option,

$$\begin{cases} \Delta S_t u + \text{Cash} \cdot \exp\left(r \cdot T\right) = \max\left(S_t u - K, 0\right) \\ \Delta S_t d + \text{Cash} \cdot \exp\left(r \cdot T\right) = \max\left(S_t d - K, 0\right) \end{cases}$$

We have two equations with two unknowns. If we subtract the second equation from the first, it is easy to see that:

$$\Delta = \frac{\max\left(S_t u - K, 0\right) - \max\left(S_t d - K, 0\right)}{S_t \left(u - d\right)}.$$

Substituting the value of Δ in the first equation, we can find that:

$$\text{Cash} = \exp\left(-r \cdot T\right) \left[\frac{\max\left(S_t d - K, 0\right) u - \max\left(S_t u - K, 0\right) d}{u - d}\right].$$

If we name V_u the option value in the case the asset is $S_0 u$ and V_d when $S_0 d$, we can obtain a general formula:

$$\Delta = \frac{V_u - V_d}{S_t \left(u - d\right)}$$

and
$$\text{Cash} = \exp(-r \cdot T) \left[\frac{V_d u - V_u d}{u - d} \right].$$

Note that we can represent the option price as a line:

$$\text{Option price} = \frac{V_u - V_d}{S_t (u - d)} \cdot S_t + \exp(-r \cdot T) \left[\frac{V_d u - V_u d}{u - d} \right]. \quad (5.7)$$

Example 5.1

Consider a European put option with a strike of 11€, maturity time 5 days. The underlying price is 10€. In the future, the underlying may reach 12 or 8€. Assume the risk-free interest rate is 0% flat. Calculate the option price?

1. Calculate the European put option's Δ.

$$\begin{aligned}
\Delta &= \frac{\text{Total change on the option price}}{\text{Total change on the asset price}} \\
&= \frac{\max(11 - 12, 0) - \max(11 - 8, 0)}{12 - 8} \\
&= \frac{0 - 3}{12 - 8} = -\frac{3}{4}.
\end{aligned}$$

2. Built a portfolio based on the European put option and $-\Delta$ times the underlying. Then, see the possible values it can take:

   - If the underlying is 12€, then the portfolio is
   $$\max(11 - 12, 0) - \left(-\frac{3}{4}\right) 12 = 0 + 9 = 9$$

   - If the underlying is 8€, then the portfolio is
   $$\max(11 - 8, 0) - \left(-\frac{3}{4}\right) 8 = 3 + 6 = 9$$

   In the two cases, we obtain the same value. The portfolio is frozen. Therefore, we can discount this amount to know how much the option is worth.

$$Option - \Delta 10 = 9 \cdot \exp\left(0\%\frac{5}{365}\right)$$

Using linear algebra

$$Option = \left(-\frac{3}{4}\right)10 + 9 \cdot \exp\left(0\%\frac{5}{365}\right) = \frac{3}{2}.$$

5.1.3 The Risk-neutral Valuation

At this point, we have seen two ways to get the option price. On the one hand, we can find the option price using replication. On the other hand, it is possible to obtain the option price by estimating some parameters. For some of us, the second one seems the most natural. We are used to extracting data from the past to forecast the future. But this is a difficult path. Not only because of the number of parameters to estimate, but also because of its complexity. For example, the risk-adjusted rate to discount the payoff.

It may feel strange that after introducing a probabilistic model, we cannot use it. However, we are going to apply a workaround. Suppose there is a parallel universe. In this universe, the assets and derivatives prices are the same as in our world. But instead, all assets must grow at the risk-free rate. This has several implications:

- In this universe, we are not aware of the risk. Or in other words, we are indifferent to risk.

- We can discount all the prices at the risk-free rate.

- The mathematical expectation of the future asset values must be the asset forward price.

- The probability of going up or down is a consequence of the fact that the future asset values must be the asset forward price.

Therefore, in this parallel universe, once we know how much an asset can go up or down, we can get the price of a derivative using mathematical expectations. Remember that the assets and derivatives prices in both universes are the same. Then, when finding the price in this universe, we also get the price in our universe.

This parallel universe is called a risk-neutral world, and our universe is the real world. The risk-neutral world is a fictional universe. Here, the assets do not have risk. The probabilities values are fictitious. It is a mathematical construction rich enough to obtain option prices. Whereas the real world is the world we live in, the assets have a premium risk, and we can forecast different growths often related to its risk level.

The first step in the risk-neutral valuation, it is obtaining the risk-neutral probabilities given a maturity time $T - t$. To do that, the asset average must be the forward value. We can simplify assuming that today is the time zero, i.e. $t = 0$, being T the distance to the maturity time. Then

$$p^* \cdot S_0 u + (1 - p^*) \cdot S_0 d = S_0 \cdot \exp(r \cdot T). \tag{5.8}$$

Doing some basic calculus, we can see that

$$p^* \cdot u - p^* \cdot d + d = \exp(r \cdot T).$$

Rearranging some factors, we observe that

$$p^*(u - d) = \exp(r \cdot T) - d.$$

Therefore, we obtain:

$$p^* = \frac{\exp(r \cdot T) - d}{(u - d)}. \tag{5.9}$$

Once we have the risk-neutral probabilities, we can obtain the average payoff and the option price will be the discounted average option price.

Example 5.2

Consider a European call option with a strike of 11€, maturity time 5 days. The underlying price is 10€. In the future, the underlying may reach 12 or 8€. Assume the risk-free interest rate is 5% flat. Calculate the option price?

To solve the problem, do the following steps:

- Calculate the risk-neutral probability:

$$p^* = \frac{\exp\left(5\% \cdot \frac{5}{365}\right) - \frac{8}{10}}{\left(\frac{12}{10} - \frac{8}{10}\right)} = 0.5017$$

- Calculate the payoff in the two scenarios.
- Calculate the mathematical expectation of the payoff:

$$\mathbb{E}\left[\text{Payoff}\right] = 0.5017 \cdot 1 + (1 - 0.5017) \cdot 0 = 0.5017.$$

- Discount the payoff from the maturity date to today:

$$\begin{aligned}\text{Option} &= \exp\left(-5\% \cdot \frac{5}{365}\right) \cdot \mathbb{E}\left[\text{Payoff}\right] \\ &= 0.999 \cdot 0.5017 = 0.5013.\end{aligned}$$

We have shown a way to price options using probabilities. But, there is a flaw in the argument of deriving the risk-neutral probability. Remember that under the risk-neutral measure the mathematical expectation of a future asset is the forward price. In equation (5.8), we have to use the forward price of a non-dividend asset. We have to use the forward price according to the underlying. Therefore, we can generalize the risk-neutral probability as

$$p^* = \frac{a - d}{(u - d)} \qquad (5.10)$$

where

- $a = \exp\left(r \cdot T\right)$ when the underlying is a non-dividend asset.
- $a = \exp\left((r - q) \cdot T\right)$ when the underlying is paying a yield dividend.
- $a = \exp\left((r - c) \cdot T\right)$ when the underlying is a currency where c is the foreign interest rate or the underlying has a yield cost c.
- $a = 1$ when the underlying is a future or forward.

The Single-Period Binomial Model ■ 67

> **Example 5.3**
>
> Consider a European call option with a strike of 11€, maturity time 5 days. The underlying price is a *future* traded at 10€. In the future, the underlying may reach 12 or 8€. Assume the risk-free interest rate is 5% flat. Calculate the option price?
> To solve the problem, do the following steps:
>
> - Calculate the risk-neutral probability:
>
> $$p^* = \frac{1 - \frac{8}{10}}{\left(\frac{12}{10} - \frac{8}{10}\right)} = 0.5$$
>
> - Calculate the payoff in the two scenarios.
>
> - Calculate the mathematical expectation of the payoff:
>
> $$\mathbb{E}\left[\text{Payoff}\right] = 0.5 \cdot 1 + (1 - 0.5) \cdot 0 = 0.5.$$
>
> - Discount the payoff from the maturity date to today:
>
> $$\text{Option} = \exp\left(-5\% \cdot \frac{5}{365}\right) \cdot \mathbb{E}\left[\text{Payoff}\right] = 0.999 \cdot 0.5 = 0.4996.$$

5.1.4 Link the Model to the Market

We have learned how to price an option using two different techniques: the replication portfolio and risk-neutral pricing. Both depend on the width of the possible values that the asset can take. Until now, we have chosen the possible values arbitrarily. But, would you be willing to pay the same price for an option with a low-risk stock underlying as for one with a high-risk stock underlying? It doesn't seem fair. After all, risk will play a role in this game. An asset with higher volatility or more uncertainty will move further. The further the asset moves, the wider the asset's price range and the more expensive the option price is. Thus, we need to relate u and d to something that depends on asset volatility or uncertainty.

To do so, we will use the following relationship

$$u = \exp\left(\sigma\sqrt{T}\right) \qquad (5.11)$$

and
$$d = \frac{1}{u} = \exp\left(-\sigma\sqrt{T}\right) \qquad (5.12)$$

where σ is the asset volatility and T is the maturity time of a one-step binomial model.

Example 5.4

Assume today's asset price $S_0 = 100$, the volatility is 30% annual, and the continuously compounded interest rate is $r = 10\%$. What is the price of a call option with strike $K = 100$ and maturity time $T = 1$ year, assuming the asset does not pay dividends and the Binomial model?

To solve the problem, do the following steps:

- Calculate u and d:

$$u = \exp\left(30\% \cdot \sqrt{1}\right), d = \exp\left(-30\% \cdot \sqrt{1}\right).$$

- Calculate the risk-neutral probability p^*

$$p^* = \frac{\exp\left(30\% \cdot 1\right) - \exp\left(-30\%\right)}{\exp\left(30\%\right) - \exp\left(-30\%\right)} = 0.5982$$

- Calculate the possible asset prices.
  In this case, it can go up by $100 \cdot \exp\left(30\%\right) = 134.98$ or down $100 \cdot \exp\left(-30\%\right) = 74.08$.

- Calculate the possible payoffs.
  The payoff can be $\max(134.98 - 100, 0) = 34.08$ or $\max(74.08 - 100, 0) = 0$.

- Calculate the mathematical expectation of the payoff:

$$\mathbb{E}\left[\text{Payoff}\right] = 0.5982 \cdot 34.08 + (1 - 0.5982) \cdot 0 = 20.93.$$

- Discount the payoff from the maturity date to today:

$$\text{Option} = \exp\left(-10\% \cdot 1\right) \cdot \mathbb{E}\left[\text{Payoff}\right] = 0.905 \cdot 20.93 = 18.938.$$

5.2 THE MULTI-PERIOD BINOMIAL MODEL

Now that we know how the binomial model works and understand risk-neutral pricing. It's time to make things more complex. Until now, all the trees we have seen had one period. That means that the price moves from today to maturity time in one step. But, we can expand the trees adding a new binomial to each previous scenario (see Figure 5.9).

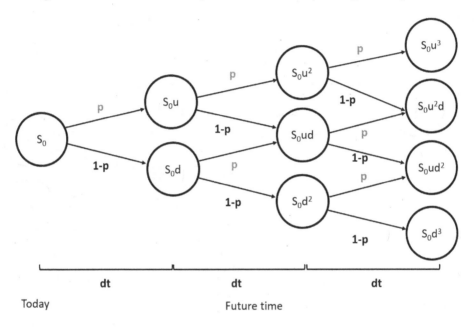

Figure 5.9 Multi-Period Binomial Model.

To do that, we break the time until maturity into small time units dt. These time units have the same size. We refer to them as time steps. At each time step, there is a new layer with an extra scenario. For example, in Figure 5.9, there are three layers. To find the size of each time step, we need to calculate:

$$dt := \frac{T - t}{\text{N}^\circ \text{ Layers}}.$$

Without losing generality, we can consider $t = 0$ having

$$dt = \frac{T}{\text{N}^\circ \text{ Layers}}.$$

Note that the number of layers is the sum of powers of u and d. And the number of layers plus one is the number of scenarios. For example, in the first layer, we have $S_0 u$ or $S_0 d$. In both cases, the sum of the powers is 1. In the second layer, we have $S_0 u^2$, $S_0 ud$, or $S_0 d^2$, then the sum of the powers is 2.

To more easily identify each scenario, we label them (see Figure 5.10).

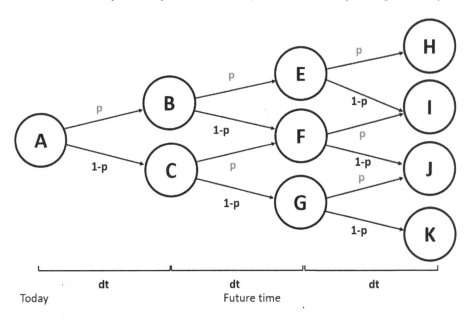

Figure 5.10 Labeling the scenarios of a Multi-Period Binomial model.

Observe that the only way to reach the scenario H is by:

$$A - B - E - H.$$

The same happens with the scenario K:

$$A - C - G - K.$$

But it is not the same for scenario J:

$$A - B - F - J,$$
$$A - C - F - J,$$
$$A - C - G - J.$$

When there are a few layers, we can count how many chances we have to end in a scenario. But when the number of layers increases, this quickly becomes more difficult. There is an easy way to know how many different paths there are to end in a scenario. Pascal's triangle which it is also known as Tartaglia's triangle.

Layer 0:					1			
Layer 1:				1		1		
Layer 2:			1		2		1	
Layer 3:		1		3		3		1
Layer 4:	1		4		6		4		1
Layer 5: 1		5		10		10		5		1

When there is one layer, there is only one path by scenario. When we increase the number of layers, we see how the possibilities quickly increase. In fact, each node of the Pascal triangle can be represented by a binomial coefficient, $Bin(n,k)$ being n the layer and k the scenario. Remember that the binomial coefficient is defined by the next expression:

$$Bin(n,k) = \binom{n}{k} = \frac{n!}{k!(n-k)!}. \tag{5.13}$$

Example 5.5

If we want to know how many paths has the second scenario of a Binomial model with 5 layer, we have to calculate $Bin(5,2)$.

$$Bin(5,2) = \binom{5}{2} = \frac{5!}{2!(5-2)!} = \frac{5 \cdot 4 \cdot 3 \cdot 2 \cdot 1}{(2 \cdot 1) \cdot (3 \cdot 2 \cdot 1)} = \frac{120}{2 \cdot 6} = 10.$$

It is interesting to observe in a bar chart the paths for each scenario. As the number of time steps increases, so does the number of scenarios. The result is a bell-shaped distribution. The distribution reminds us of a normal distribution. This is because, in the limit, the binomial distribution tends to be a normal distribution (see Figure 5.11).

72 ■ The Binomial Model

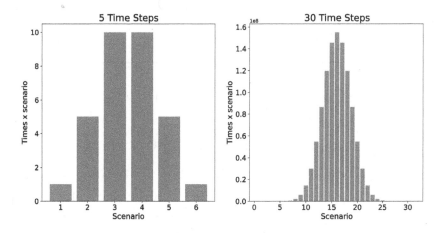

Figure 5.11 How many times a scenario is reached depending on the time steps.

5.2.1 Adjusting the Parameters

We have seen that the Multi-Period Binomial model is richer than the Single-Period version. If the time step is small enough, the amount of prices we simulate is related to a normal distribution. But, we have to adapt the parameters to this model.

To calculate the average underlying price, we can use the binomial coefficients. When there is only one time step (or layer), we have:

$$p^* \cdot S_0 u + (1 - p^*) \cdot S_0 d.$$

When there are two time steps (or layers):

$$(p^*)^2 \cdot S_0 u^2 + 2 \cdot p^* \cdot (1 - p^*) \cdot S_0 ud + (1 - p^*)^2 \cdot S_0 d^2.$$

When there are three time steps (or layers):

$$(p^*)^3 \cdot S_0 u^3 + 3 \cdot (p^*)^2 \cdot (1 - p^*) \cdot S_0 u^2 d$$
$$+ 3 \cdot p^* \cdot (1 - p^*)^2 \cdot S_0 ud^2 + (1 - p^*)^3 \cdot S_0 d^3.$$

We can generalize for n time steps by:

$$S_0 \sum_{k=0}^{n} \binom{n}{k} (p^* \cdot u)^{n-k} \cdot ((1 - p^*) \cdot d)^k. \tag{5.14}$$

Using the Binomial Theorem, it is easy to see that Equation (5.4) can be written as:

$$S_0 \Big(p^* \cdot u + (1-p^*) \cdot d\Big)^n.$$

Remember that under the risk-neutral valuation, the average underlying price is the forward price:

$$S_0 \Big(p^* \cdot u + (1-p^*) \cdot d\Big)^n = S_0 \cdot \exp(r \cdot (T-t))$$

Then

$$\Big(p^* \cdot u + (1-p^*) \cdot d\Big) = \exp\Big(r \cdot \frac{(T-t)}{n}\Big).$$

Using the definition of time step, it is easy to see that

$$p^*(u-d) = \exp(r \cdot dt) - d.$$

Therefore,

$$p^* = \frac{\exp(r \cdot dt) - d}{u - d}. \qquad (5.15)$$

We have used the forward price of an underlying without dividends, but as we did it in Section 5.1.3, we can generalize by using

$$p^* = \frac{a - d}{(u - d)} \qquad (5.16)$$

where

- $a = \exp(r \cdot dt)$ when the underlying is a non-dividend asset.
- $a = \exp((r - q) \cdot dt)$ when the underlying is paying a yield dividend.
- $a = \exp((r - c) \cdot dt)$ when the underlying is a currency where c is the foreign interest rate or the underlying has a yield cost c.
- $a = 1$ when the underlying is a future or forward.

74 ■ The Binomial Model

It is also necessary to adapt the definitions of u and d having

$$u = \exp\left(\sigma\sqrt{dt}\right), \tag{5.17}$$

$$d = \frac{1}{u} = \exp\left(-\sigma\sqrt{dt}\right) \tag{5.18}$$

In this case, we have that $u \cdot d = 1$, then the tree can be simplified to the following form (see Figure 5.12).

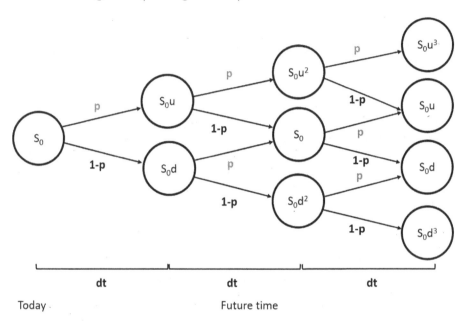

Figure 5.12 Multi-Period Binomial Model. Simplifying the notation.

5.2.2 Pricing a European Option

Now that we know what the parameters are like to price an option in a Multi-Period model, we show how to price an option. There are two ways to price a European Option with the Binomial model. We will review the extended scheme. This scheme will be convenient to us when we need to price an American option.

5.2.2.1 Extended Framework

Given a European call option on an underlying asset without dividends with a strike price of 10€ and a maturity of 1 year. The underlying price is 10€ and has a volatility of 20% and the interest rates are 5%. We want to calculate the fair price of an option using a Binomial model with 3 time steps. To do that, we need to do the following steps:

1. Calculate the time step size.

$$dt = \frac{T-t}{3} = \frac{1-0}{3} = 0.33 \text{ years.}$$

2. Calculate u and d.

$$u = \exp\left(20\% \cdot \frac{1}{3}\right) = 1.1224,$$

$$d = \exp\left(-20\% \cdot \frac{1}{3}\right) = 0.8909.$$

3. Calculate the risk-neutral probability p^*.

$$p^* = \frac{\exp(r \cdot dt) - d}{(u-d)} = \frac{\exp(5\% \cdot \frac{1}{3}) - d}{(u-d)} = 0.5437.$$

4. Populate the price tree (see Figure 5.13).

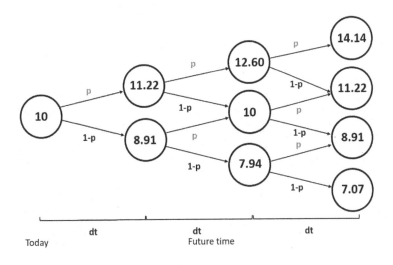

Figure 5.13 Populating the tree prices.

For example, the node C is $S_0 d = 8.91$ and the node H is $S_0 u^3 = 14.14$.

76 ■ The Binomial Model

5. Calculate the option payoff at the last node (see Figure 5.14).

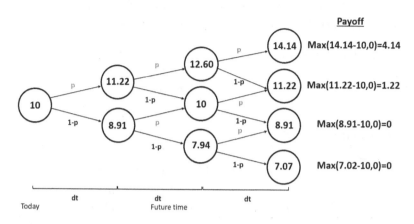

Figure 5.14 Finding the payoff.

Note that in this case, we have a call option. We should use the corresponding Payoff function.

6. Calculate the option price at nodes E, F, and G as the discounted expectation (see Figure 5.15).

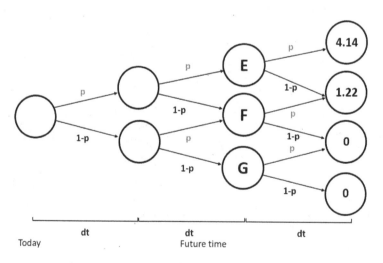

Figure 5.15 Calculating backwards the first layer.

Node E: $\exp\left(-r \cdot \dfrac{1}{3}\right) \cdot \Big[p^* \cdot \max(14.14 - 10, 0)$
$- (1 - p^*) \cdot \max(11.22 - 10, 0)\Big] = 2.76$

Node F: $\exp\left(-r \cdot \dfrac{1}{3}\right) \cdot \Big[p^* \cdot \max(11.22 - 10, 0)$
$- (1 - p^*) \cdot \max(8.91 - 10, 0)\Big] = 0.65$

Node G: $\exp\left(-r \cdot \dfrac{1}{3}\right) \cdot \Big[p^* \cdot \max(8.91 - 10, 0)$
$- (1 - p^*) \cdot \max(7.07 - 10, 0)\Big] = 0$

7. Calculate the option price at nodes B and C as the discounted expectation (see Figure 5.16).

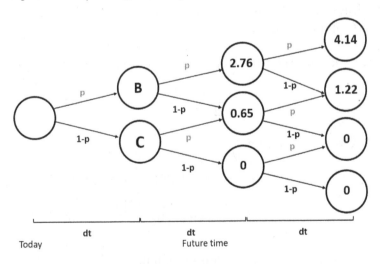

Figure 5.16 Calculating backwards the second layer.

Node B: $\exp\left(-r \cdot \dfrac{1}{3}\right) \cdot \Big[p^* \cdot 2.76 - (1 - p^*) \cdot 0.65\Big] = 1.77$

Node C: $\exp\left(-r \cdot \dfrac{1}{3}\right) \cdot \left[p^* \cdot 0.65 - (1-p^*) \cdot 0\right] = 0.35$

8. Repeat for node A.

Node A: $\exp\left(-r \cdot \dfrac{1}{3}\right) \cdot \left[p^* \cdot 1.77 - (1-p^*) \cdot 0.35\right] = 1.10$

9. The price of the option is 1.10.

First, we move through the nodes by finding the possible values of the underlying. Once we reach the time of maturity, we regress. We calculate the expectation of two adjacent scenarios and discount the price to the nearest time step (or layer). Then, we proceed iteratively until today. Once the framework is clear, the calculus involved is quite simple.

Example 5.6

Consider a European Put Option. The USDEUR rate is at 0.9 € and has a volatility of 10%. Assume interest rates are flat, with a US risk-free rate of 7% and an EU risk-free rate of 4%. Using a two-step Binomial tree, what is the price of an option with a strike price of 0.8€ and a maturity of 1 year?

1. Calculate the time step size.

$$dt = \frac{T-t}{2} = \frac{1-0}{2} = 0.5 \text{ years.}$$

2. Calculate u and d.

$$u = \exp\left(10\% \cdot \frac{1}{2}\right) = 1.073,$$

$$d = \exp\left(-10\% \cdot \frac{1}{2}\right) = 0.932.$$

3. Calculate the risk-neutral probability p^*.

$$p^* = \frac{\exp\left((r_{EU} - r_{US}) \cdot dt\right) - d}{(u-d)}$$

$$= \frac{\exp\left(-3\% \cdot \frac{1}{2}\right) - d}{(u-d)} = 0.377.$$

4. Populate the price tree (see Figure 5.17).

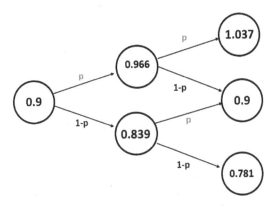

Figure 5.17 Populating the tree prices with a FX asset.

For example, the node C is $S_0 d = 0.839$ and the node D is $S_0 u^2 = 1.037$.

5. Calculate the option payoff at the last node (see Figure 5.18).

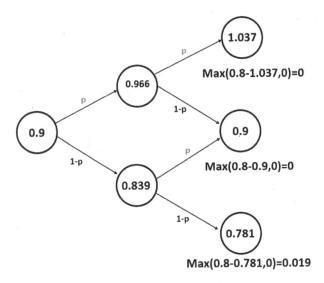

Figure 5.18 Finding the payoff with a FX asset.

Note that in this case, we have a put option. We should use the corresponding Payoff function.

6. Calculate the option price at nodes B and C as the discounted expectation (see Figure 5.19).

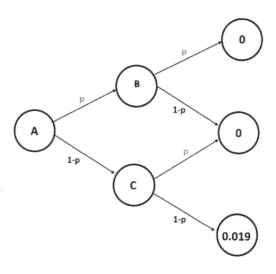

Figure 5.19 Calculating backwards the first layer with a FX asset.

Node B: $\exp\left(-r_{EU} \cdot \dfrac{1}{2}\right) \cdot \Big[p^* \cdot \max(0.8 - 1.037, 0)$

$$- (1 - p^*) \cdot \max(0.8 - 0.9, 0)\Big] = 0$$

Node C: $\exp\left(-r_{EU} \cdot \dfrac{1}{2}\right) \cdot \Big[p^* \cdot \max(0.8 - 0.9, 0)$

$$- (1 - p^*) \cdot \max(0.8 - 0.781)\Big] = 0.0114$$

7. Repeat for node A.

$$\text{Node A: } \exp\left(-r_{EU} \cdot \frac{1}{2}\right) \cdot \left[p^* \cdot 0 - (1-p^*) \cdot 0.0114\right] = 0.007$$

The price of the option is 0.007.

Example 5.7

Consider a European Put Option over a stock with price 5€, a dividend yield of 3% and has a volatility of 20%. Assume the interest rates are flat with a rate of 2%. Using a two-step Binomial tree, what is the price of an option with a strike price of 8€ and a maturity of 2 year?

1. Calculate the time step size.

$$dt = \frac{T-t}{2} = \frac{2-0}{2} = 1 \text{ year.}$$

2. Calculate u and d.

$$u = \exp(20\% \cdot 1) = 1.221,$$
$$d = \exp(-20\% \cdot 1) = 0.819.$$

3. Calculate the risk-neutral probability p^*.

$$p^* = \frac{\exp((r-y\%) \cdot dt) - d}{(u-d)}$$
$$= \frac{\exp(-1\% \cdot 1) - d}{(u-d)} = 0.425.$$

4. Populate the price tree (see Figure 5.20).

The Binomial Model

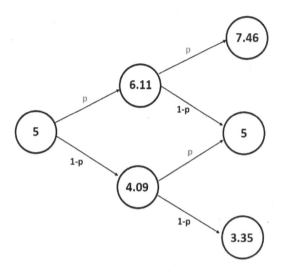

Figure 5.20 Populating the tree prices when the asset pays dividends.

For example, the node C is $S_0 d = 4.09$ and the node F is $S_0 d^2 = 3.35$.

5. Calculate the option payoff at the last node (see Figure 5.21).

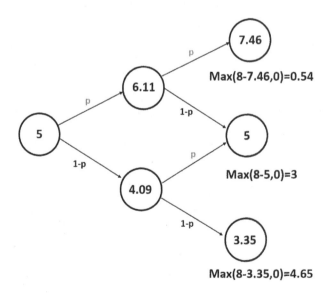

Figure 5.21 Finding the payoff when the asset pays dividends.

Note that in this case, we have a put option. We should use the corresponding Payoff function.

6. Calculate the option price at nodes B and C as the discounted expectation (see Figure 5.22).

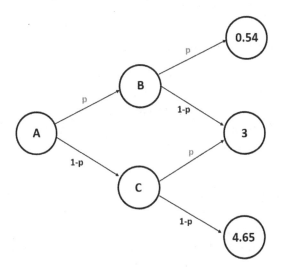

Figure 5.22 Calculating backwards the first layer the asset pays dividends.

Node B: $\exp(-r \cdot 1) \cdot \Big[p^* \cdot \max(8 - 7.46, 0)$

$$- (1 - p^*) \cdot \max(8 - 5, 0) \Big] = 1.915$$

Node C: $\exp(-r \cdot 1) \cdot \Big[p^* \cdot \max(8 - 5, 0)$

$$- (1 - p^*) \cdot \max(8 - 3.35, 0) \Big] = 3.869$$

> 7. Repeat for node A.
>
> $$\text{Node A: } \exp(-r \cdot 1) \cdot \left[p^* \cdot 1.915 - (1-p^*) \cdot 3.869\right] = 2.977$$
>
> The price of the option is 2.977.

5.2.2.2 Simplified Framework

The European option depends solely on the underlying values at the maturity time. The price of an option is the discounted expected value. Therefore, the main problem is to calculate that expected value. We can use the properties of the Binomial coefficients and the possible payoff values to find the price of an European option.

In the example of the previous section, we have that the option price is given by

$$\exp(-5\% \cdot 1) \left[(p^*)^3 \cdot 4.14 + 3(p^*)^2(1-p^*) \cdot 1.22 \right.$$
$$\left. + 3(p^*)(1-p^*)^2 \cdot 0 + (1-p^*)^3 \cdot 0\right]$$
$$= \exp(-5\% \cdot 1)[1.1610]$$
$$= 1.1043.$$

In fact, we have the following formula

$$\exp(-r \cdot T) \left[\sum_{k=0}^{n} \binom{n}{k}(p^*)^{n-k}(1-p^*)^k \max(S_0 u^{n-k} d^k - K, 0)\right].$$

An analogous formula can be found out for the European put options.

Although this method is more direct, it can only be used for options that depend on the terminal value.

5.2.3 Early Exercise

We have learned how to price European options. Unfortunately, it is more difficult when pricing early exercise options such as American options. In this case, we have to choose at each moment, i.e., $t < T$, if it is better to keep the option or execute it. That means that to calculate the price of an American option, we have to look at each node to see if it is higher

- The corresponding European option price.
- The intrinsic value.

There is only one exception. When **the underlying is a stock that does not pay dividends**, some non-trivial computations (by making use of advanced mathematics) show that the price of an American call is equal to the price of the corresponding European call.

Recovering the Example 5.6, we can see how the pricing changes.

Example 5.8

The first 6 steps are the same. Asset prices tree is in Figure 5.17 and the European option tree is given by Figure 5.23:

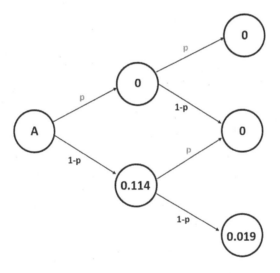

Figure 5.23 Considering the American option.

Now, we have to add a new step. Compare the option price and the intrinsic value. In particular, for the nodes B and C.

In B, we have intrinsic value: $\max(0.80 - 0.966, 0) = 0$. So, in this case, the intrinsic value and the option value are the same, 0.

In C, we have the intrinsic value: $\max(0.80 - 0.839, 0) = 0$. So, in this case, the option value of 0.114 is greater than the intrinsic value of 0. Then, we keep the biggest one, 0.114.

86 ■ The Binomial Model

With the maximum values, we calculate the value for node A. In this case, is the same as Example 5.6 because it was not optimal to execute it.

Node A: $\exp\left(-r_{EU} \cdot \dfrac{1}{2}\right) \cdot \Big[p^* \cdot \max(0,0)$

$- (1-p^*) \cdot \max(0.0114, 0)\Big] = 0.007$

Now, we add a new step. Compare the option price and the intrinsic value. In A, the intrinsic value is $\max(0.80 - 0.9, 0) = 0$. Then the option price is bigger than the intrinsic value, so the option price is 0.0007.

This option is not optimal to execute before maturity time.

We repeat the process with the Example 5.7.

Example 5.9

The first 6 steps are the same. The prices tree is in Figure 5.20 and the option tree is as in Figure 5.20.

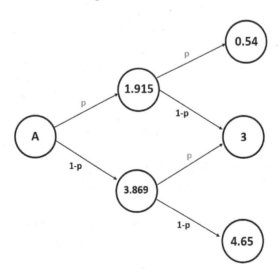

Figure 5.24 Considering the American option when the asset pays dividends.

Now, we have to add a new step. Compare the option price and the intrinsic value. In particular, for the nodes B and C.

In B, we have intrinsic value: $\max(8 - 6.11, 0) = 1.89$. So, in this case, the option value is bigger than the intrinsic value. Then, we choose the option value because it is not optimal to execute the option here.

In C, we have the intrinsic value: $\max(8 - 4.09, 0) = 3.91$. So, in this case, the intrinsic value is higher than the option value. It is optimal to execute the option here. So, we choose the intrinsic values.

With the maximum values, we calculate the value for node A. In this case, it is different than Example 5.7 because there is a case when it is optimal to execute it before maturity.

$$\text{Node A: } \exp(-r \cdot 1) \cdot \Big[p^* \cdot \max(1.915, 1.89)$$
$$- (1 - p^*) \cdot \max(3.869, 3.91) \Big] = 3$$

Now, we add a new step. Compare the option price and the intrinsic value at the initial node. In A, the intrinsic value is $\max(8 - 5, 0) = 3$. Then the option price is the same as the intrinsic value.

The price of the option is 3.

5.3 THE GREEKS IN THE BINOMIAL MODEL

In the financial sector, to understand and control the movement of assets and their risk, we are interested in knowing the behavior of the price with respect to its parameters. For example, how much an option can change in value when the stock falls or how much the option can change in value when there is an increase in volatility. These types of sensitivities can help us to understand why the price has changed from one day to another or quantify the possible losses or profits in a specific scenario. To do so, we have a powerful mathematical tool, the derivatives. Remember that a derivative is how much a function changes when the parameter changes. In mathematical terms, given a differentiable function f, we can write the derivative as

$$f'(x) = \lim_{h \to 0} \frac{f(x+h) - f(x)}{h}.$$

In graphical terms, we can observe this behavior in Figure 5.25:

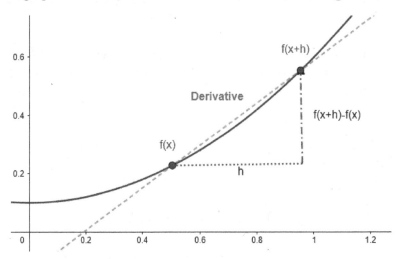

Figure 5.25 Visualize the derivative concept.

There are two interesting reasons for knowing the derivatives of a function, both related. First, understand the function's dependency with respect to a variable. The second is to use that relationship to construct a function approximation using the Taylor series. The Taylor series approximates a function with a series expansion. The only requirement is that the function must be differentiable at the approximation point.

For example, given a differentiable function on the point x_0. If we know the values of $f(x_0), f'(x_0), f''(x_0)$, and $f'''(x_0)$, we can approximate the function by

$$f(x) = f(x_0) + f'(x_0)(x - x_0) + \frac{1}{2}f''(x_0)(x - x_0)^2 + \frac{1}{3!}f'''(x_0)(x - x_0)^3.$$

In general, given a function infinitely differentiable, we have the series expansion

$$\begin{aligned}f(x) &= f(x_0) + f'(x_0)(x - x_0) + \frac{1}{2}f''(x_0)(x - x_0)^2 \\ &+ \frac{1}{3!}f'''(x_0)(x - x_0)^3 + \frac{1}{4!}f^{(4)}(x - x_0)^4 + \ldots\end{aligned}$$

Written on a compact form

$$f(x) = \sum_{n=0}^{\infty} \frac{1}{n!} f^{(n)}(x_0)(x-x_0)^n. \qquad (5.19)$$

Example 5.10

Write the Taylor series approximation of order 3 for the function $f(x) = \exp(x)$ at the point $x_0 = 0$.

We know that all the derivatives of $f(x) = \exp(x)$ are the same. It is $f^{(n)}(x) = \exp(x)$.

If we evaluate the function and the derivatives at $x_0 = 0$, we have that $f^n(x_0) = \exp(0) = 1$.

Then,

$$\begin{aligned} \exp(x) &\approx 1 + (x-0) + \frac{1}{2}(x-0)^2 + \frac{1}{3!}(x-0)^3 \\ &= 1 + x + \frac{1}{2}x^2 + \frac{1}{6}x^3. \end{aligned}$$

In graphical terms, we can observe this behavior in Figure 5.26:

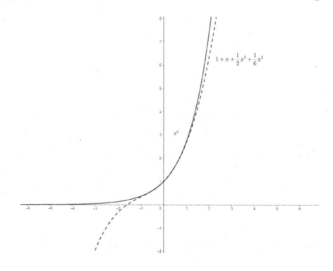

Figure 5.26 Taylor approximation of the exponential.

90 ■ The Binomial Model

> We note that locally, near the approximate point $x_0 = 0$, the original function and the approximation are very similar. However, it does not occur in the whole domain. In other words, we cannot find a global approximation. Adding more terms from the Taylor series will improve the comparison. In this case, we will need infinite terms in the approximation to adjust to the function.

In our case, the function is the option price and we will refer as V.

5.3.1 Delta

We have seen that Delta, Δ, is how many units of the underlying we need to sell to hedge our position. By definition, Δ is the derivative respect the underlying. In the Binomial model, we can calculate it as

$$\Delta = \frac{V^+ - V^-}{(u-v)S_0}$$

where V^+ is the option payoff at node B and V^- at node C. In the limit, as the time step approaches zero, the delta becomes $\Delta = \frac{\partial V}{\partial S}$.

Example 5.11

In the Example 5.6, we have that $V^+ = 0$ and $V^- = 0.114$. Also, $uS_0 = 0.966$ and $vS_0 = 0.839$. Then,

$$\Delta = \frac{V^+ - V^-}{(u-d)S_0} = \frac{0 - 0.114}{0.966 - 0.839} = -0.898.$$

5.3.2 Gamma

The Gamma, Γ, it is also a derivative respect to the underlying. It is the convexity of the option price with respect to the underlying or the derivative respect the delta of the option. In Physics, when studying the equations of motion, the first derivative represents the speed and the second the acceleration. In the limit, as the time step approaches zero, the delta becomes $\Gamma = \frac{\partial^2 V}{\partial S^2}$.

In the Binomial model, we can calculate it as

$$\Gamma = \frac{\Delta^+ - \Delta^-}{(u-d)S_0}$$

where Δ^+ is Δ at node B and Δ^- is Δ at node C.

Example 5.12

In the Example 5.6, we have that

$$\Delta^+ = \frac{V^{++} - V^{+-}}{(u^2 - 1)S_0} = \frac{0 - 0}{1.037 - 0.9} = 0$$

where V^{++} is the option price at node D and V^{+-} is the option price at node E. Also

$$\Delta^- = \frac{V^{+-} - V^{--}}{(1 - d^2)S_0} = \frac{0 - 0.114}{0.9 - 0.781} = -0.958$$

where V^{--} is the option price at node F. Now,

$$\Gamma = \frac{0 - (-0.958)}{0.966 - 0.839} = 7.54.$$

5.3.3 Theta

The Theta, Θ, is the option sensitivity to time. How much value will change when a day pass.

$$\Theta = \frac{\partial V}{\partial t}$$

In the Binomial model, we define Θ as

$$\Theta = \frac{\frac{1}{2}(V^+ + V^-) - V}{dt}.$$

where dt is a time step.

Example 5.13

In the Example 5.6, we have that

$$\Theta = \frac{\frac{1}{2}(0 - 0.114) - 0.007}{0.5} = -0.128.$$

5.3.4 Vega

One of the most important drivers for pricing the options is volatility. The vega, ϑ, is the sensitivity with respect to the volatility. When the time step goes to zero, ϑ becomes

$$\vartheta = \frac{\partial V}{\partial \sigma}.$$

In this case, we cannot use the node values to calculate the sensitivity. We have to calculate the option price twice. The first time, we will use the volatility $\sigma + h$, and the second time, we will use $\sigma - h$ with h a small number. Then, we apply a slight modification of the derivative

$$\vartheta = \frac{V(\sigma + h) - V(\sigma - h)}{(\sigma + h) - (\sigma - h)} = \frac{V(\sigma + h) - V(\sigma - h)}{2h}.$$

This version of the derivative is the central difference approximation where the value we are interested in is between the two points we use to do the calculation. In practice, usually h is 1%, but it can be others such as 0.01%, 0.1%.

5.3.5 Rho

The last variable is the risk-free interest rates. The rho, ρ, is the sensitivity with respect to the risk-free interest rates. When the time step goes to zero, ρ becomes

$$\rho = \frac{\partial V}{\partial r}.$$

As we have seen with vega, we cannot use the node values to calculate it. Similarly, we can calculate the sensitivity by

$$\rho = \frac{V(r + h) - V(r - h)}{(r + h) - (r - h)} = \frac{V(r + h) - V(r - h)}{2h}.$$

5.3.6 Approximating the Price Function

Once we know the price of an option and its Greeks, we can approximate the price with respect to its variables. In this case, we have to apply it in a multivariate setting. Then, we have

$$\begin{aligned}V(S, T, r, \sigma) &\approx V(S_0, T_0, r_0, \sigma_0) + \Delta \cdot (S - S_0) + \tfrac{1}{2}\Gamma \cdot (S - S_0)^2 \\ &\quad + \theta \cdot (T - T_0) + \vartheta \cdot (\sigma - \sigma_0)\end{aligned}$$

where S_0, T_0, r_0, and σ_0 are the values observed now in the market.

This approximation is widely used in the financial market. It allows us to segregate the market movements into different risk factors. Moreover, it explains why a derivative price has changed from one date to another. Altogether, it can help define whether it is necessary to have a hedging policy and quantify the possible risks.

5.4 CODING THE BINOMIAL MODEL

In this chapter, we have learned how to price an option using the binomial model, including some features such as early exercise. In practice, we need to find a way to find these prices quickly. On the one hand, we will need to increase the number of time steps to gain precision in our prices. On the other hand, we may be interested in calculating different sensitivities by calculating the option prices with different parameters. The best way to obtain this efficiency is by encoding the binomial model. To this end, we need the computer to follow the same steps that we do by hand.

The first problem that arises when writing code, especially for beginners, is the blank page syndrome. To avoid that, and for many other advantages, the best thing is to plan what you want to code, which are the functionalities you want, where the data comes from, what you expect the user does, ... and do a scheme. This scheme is known as pseudocode, a plain language description of the steps in an algorithm.

In our case, we want to write a function that calculates the Binomial model given some parameters. We don't expect the user to introduce it. Therefore, we have to repeat the same scheme as we did when pricing

1. Calculate u and d.
2. Calculate the risk-neutral probability.
3. Create a tree with future prices.
4. Calculate backward the option value.

The first two points are easy to reproduce by code. It is only necessary to write the formula in generic form.

```
#We upload the maths library
import numpy as np
```

94 ■ The Binomial Model

```
def get_updown_probabilities(r,sigma,T,n):
#This function calculate the up, down ratio
#and the risk-neutral probability

#Inputs:
# r: risk-free rate
# sigma: volatility
# T: Maturity time
# n: number of steps

    # We calculate the time step size
    dt=T/n

    #We apply the u and d formulation
    u=np.exp(sigma*np.sqrt(dt))
    d=1/u

    #Then, we calculate the risk-neutral probability
    p=(np.exp(r*dt) -d)/(u-d)

    return (u,d,p)
```

The first challenge is how to reproduce the tree by code. If we want to keep all the price states, we can store them in a matrix. However, it is difficult to control which state we are in. To make it easier, we are going to reshape the tree (see Figure 5.27):

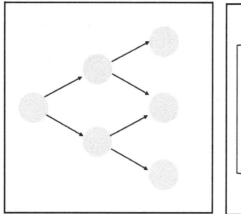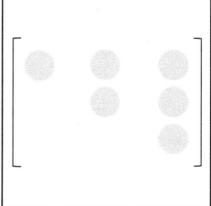

Figure 5.27 Changing the shape.

We have changed the figure of the classical tree for an upper triangular matrix. We need an algorithm to move through all the matrix positions.

```
#Algorithm to move through all positions of a matrix

#We upload the maths library
import numpy as np

#We create a Matrix with 4 positions
Matrix=np.zeros((4,4))

#We create a variable to number all the matrix positions
count=1

#For each column
for j in range(0,4):
    #For each row
    for i in range(0,4):
        Matrix[i,j]=count
        count=count+1

#Print Matrix
Matrix
```

```
array([[ 1.,  5.,  9., 13.],
       [ 2.,  6., 10., 14.],
       [ 3.,  7., 11., 15.],
       [ 4.,  8., 12., 16.]])
```

But, we will not use the values below the diagonal. We need to add more conditions. In the simplest way, we have the following modification.

```
#Algorithm to move through all positions of a matrix

#We upload the maths library
import numpy as np

#We create a Matrix with 4 positions
Matrix=np.zeros((4,4))

#We create a variable to number all the matrix positions
count=1

#For each column
for j in range(0,4):
    #For each row
    for i in range(0,4):
        #We add a condition to move only in the upper-diagonal matrix
```

96 ■ The Binomial Model

```
        if j>=i:
            Matrix[i,j]=count
            count=count+1

#Print Matrix
Matrix
```

```
array([[  1.,   2.,   4.,   7.],
       [  0.,   3.,   5.,   8.],
       [  0.,   0.,   6.,   9.],
       [  0.,   0.,   0.,  10.]])
```

We can be more elegant, including the position condition into the second for.

```
#Algorithm to move through all positions of a matrix

#We upload the maths library
import numpy as np

#We create a Matrix with 4 positions
Matrix=np.zeros((4,4))

#We create a variable to number all the matrix positions
count=1

#For each column
for j in range(0,4):
    #For each row
    for i in range(0,j+1):
        Matrix[i,j]=count
        count=count+1

#Print Matrix
Matrix
```

```
array([[  1.,   2.,   4.,   7.],
       [  0.,   3.,   5.,   8.],
       [  0.,   0.,   6.,   9.],
       [  0.,   0.,   0.,  10.]])
```

Let's see the path where the asset grows the most, formed by the nodes: $S_0 u^0 d^0$, $S_0 u^1 d^0$, $S_0 u^2 d^0$, and $S_0 u^3 d^0$. The exponent of u is equal to the column number starting from zero. In other words, when we move one column to the right but stay on the same line, we raise the exponent of u by one degree (see Figure 5.28).

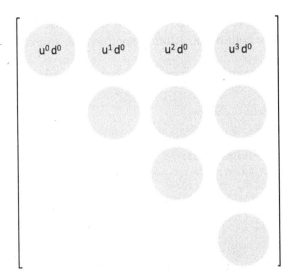

Figure 5.28 How the always raise scenario fills the matrix.

Now let's see the last column, formed by the nodes: $S_0 u^3 d^0$, $S_0 u^2 d^1$, $S_0 u^1 d^2$, and $S_0 u^0 d^3$. In the first row, the exponent of u is the column number (starting at zero). Every time we go down a row, the exponent of u decreases by one unit, and the exponent of d increases by one unit (see Figure 5.29).

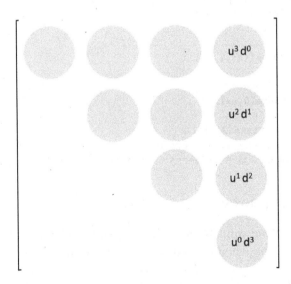

Figure 5.29 How the payoff scenarios fills the matrix.

98 ■ The Binomial Model

In the for loop, the variable i control the row position and the variable j control the column position. Therefore, we can adapt the code as follows.

```python
#Function to obtain prices on a Binomial model

#We upload the maths library
import numpy as np

def get_prices(S,r,sigma,T,n):
#This function calculate the up, down ratio
#and the risk-neutral probability

#Inputs:
#r: risk-free rate
#sigma: volatility
#T: Maturity time
#n: number of steps


    # We calculate the time step size
    dt=T/n

    #We apply the u and d formulation
    u=np.exp(sigma*np.sqrt(dt))
    d=1/u

    #Then, we calculate the risk-neutral probability
    p=(np.exp(r*dt) -d)/(u-d)

    #We create a matrix to store the prices
    Stree=np.zeros((n+1, n+1))

    #For each time step
    for j in range(0,n+1):
        #For each path
        for i in range(0, j+1):
            #The nodes are powers of u and d
            Stree[i,j] = S*(u**(j-i))*(d**(i));

    return (u,d,p,Stree)
```

Note that if we want to do 2 time steps, we will have three columns: the initial position, the first time step, and the second one. Now, we will create a new matrix with the intrinsic option value. At maturity time, that is the last column, the intrinsic value corresponds to the payoff value. We need to create two new variables: the strike price and the option type, to apply the corresponding payoff function.

```python
#Function to obtain prices on a Binomial model
#As well as the Intrinsic option values

#We upload the maths library
import numpy as np

def get_prices_and_intrinsic(S,K,r,sigma,T,n,otype):
#This function calculate the up, down ratio
#and the risk-neutral probability

#Inputs:
#r: risk-free rate
#sigma: volatility
#T: Maturity time
#n: number of steps

    # We calculate the time step size
    dt=T/n

    #We apply the u and d formulation
    u=np.exp(sigma*np.sqrt(dt))
    d=1/u

    #Then, we calculate the risk-neutral probability
    p=(np.exp(r*dt) -d)/(u-d)

    #We create a matrix to store the prices
    Stree=np.zeros((n+1, n+1))

    #We create a matrix to store the intrinic value
    Intrinsic=np.zeros((n+1, n+1))

    #For each time step
    for j in range(0,n+1):
        #For each path
        for i in range(0,j+1):
            #The nodes are powers of u and d
            Stree[i,j] = S*(u**(j-i))*(d**(i))

            #Depending if is a call or a put
            #we apply a different payoff function
            if otype=="call":
                Intrinsic[i,j]=np.maximum(Stree[i,j]-K,0)
            elif otype=="put":
                Intrinsic[i,j]=np.maximum(K-Stree[i,j],0)
```

```
        else:
            print("Wrong option type. Please write call or put")
return (u,d,p,Stree,Intrinsic)
```

Now, we need to calculate the option price. We calculate the option value by going backward from the maturity to the initial time. That means we need to use the same algorithm type but now starting at the penultimate column up to the first. We can adapt the second 'for' clause to specify the initial and last position and change the direction. Looking back to an example of numbering the matrix positions.

```
#Algorithm to move through all positions of a matrix

#We upload the maths library
import numpy as np

#We create a Matrix with 4 positions
Matrix=np.zeros((4,4))

#We create a variable to number all the matrix positions
count=1

for j in range(3,0,-1):
    for i in range(0,j+1):
        Matrix[i,j]=count
        count=count+1

#Print Matrix
Matrix
```

```
array([[0., 8., 5., 1.],
       [0., 9., 6., 2.],
       [0., 0., 7., 3.],
       [0., 0., 0., 4.]])
```

We want to apply this technique to move backward and to calculate the discounted option expectation (see Figure 5.30).

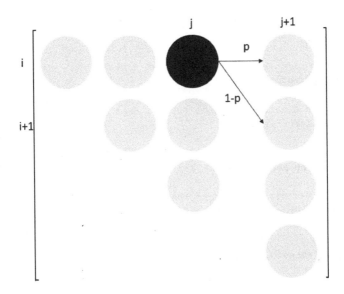

Figure 5.30 Algorithm for the Backwards calculation.

When calculating discounted option expectation of the node (i,j), we need to access to the nodes (i, j+1) with a probability p and to the node (i+1,j+1) with a probability $1 - p$. Adding that in our code, we can calculate the option value.

```
#Function to obtain prices on a Binomial model
#As well as the Intrinsic option values

#We upload the maths library
import numpy as np

def EuBinomial(S,K,r,sigma,T,n,otype):
#This function calculate the up, down ratio
#and the risk-neutral probability

#Inputs:
#r: risk-free rate
#sigma: volatility
#T: Maturity time
#n: number of steps

    # We calculate the time step size
    dt=T/n

    #We apply the u and d formulation
```

```python
u=np.exp(sigma*np.sqrt(dt))
d=1/u

#Then, we calculate the risk-neutral probability
p=(np.exp(r*dt) -d)/(u-d)

#We create a matrix to store the prices
Stree=np.zeros((n+1, n+1))

#We create a matrix to store the intrinic value
Intrinsic=np.zeros((n+1, n+1))

#We create a matrix to store the option value
Option=np.zeros((n+1, n+1))

#For each path
for j in range(0,n+1):
    #For each time step
    for i in range(0,j+1):
        #The nodes are powers of u and d
        Stree[i,j] = S*(u**(j-i))*(d**(i))

        #Depending if is a call or a put
        #we apply a different payoff function
        if otype=="call":
            Intrinsic[i,j]=np.maximum(Stree[i,j]-K,0)
        elif otype=="put":
            Intrinsic[i,j]=np.maximum(K-Stree[i,j],0)
        else:
            print("Wrong option type. Please write call or put")


#For each path
for j in range(n,-1,-1):
    #For each time step
    for i in range(0, j+1):
        if j==n:
            Option[i,j] = Intrinsic[i,j]
        else:
            Option[i,j] = np.exp(-r*dt)*(Option[i,j+1]*p\
                                  + Option[i+1,j+1]*(1-p))

print("The Option price is", Option [0,0])

return (u,d,p,Stree, Intrinsic, Option)
```

The code returns all the variables and writes on the screen the option price. That will come in handy to help us practice our hand calculations.

But we can change the return line to choose which variables we want to return. The programmer has to think about what inputs or outputs the code needs, but also how the code will obtain them. For example, we can ask the user an input value.

```
EuBinomial(5,8,0.02, 0.20,2,2, "put")
```

```
The Option price is 2.686315513218586
(1.2214027581601699,
 0.8187307530779818,
 0.5003342283694454,
 array([[5.        , 6.10701379, 7.45912349],
        [0.        , 4.09365377, 5.        ],
        [0.        , 0.        , 3.35160023]]),
 array([[3.        , 1.89298621, 0.54087651],
        [0.        , 3.90634623, 3.        ],
        [0.        , 0.        , 4.64839977]]),
 array([[2.68631551, 1.7345756 , 0.54087651],
        [0.        , 3.74793562, 3.        ],
        [0.        , 0.        , 4.64839977]]))
```

We can go one step further. Remember that the difference between an American option and a European option is the execution time. We try to execute the option at the best time, which is not necessarily the expiration time. To do this, we compare the price between the intrinsic value and the option value for each node. With a slight variation in the code, we can add that functionality.

```
#Function to obtain prices on a Binomial model
#As well as the Intrinsic option values

#We upload the maths library
import numpy as np

def Binomial(S,K,r,sigma,T,n,otype, EuOrAm):
#This function calculate the up, down ratio
#and the risk-neutral probability

#Inputs:
#r: risk-free rate
#sigma: volatility
#T: Maturity time
#n: number of steps


    # We calculate the time step size
    dt=T/n
```

```python
#We apply the u and d formulation
u=np.exp(sigma*np.sqrt(dt))
d=1/u

#Then, we calculate the risk-neutral probability
p=(np.exp(r*dt) -d)/(u-d)

#We create a matrix to store the prices
Stree=np.zeros((n+1, n+1))

#We create a matrix to store the intrinic value
Intrinsic=np.zeros((n+1, n+1))

#We create a matrix to store the option value
Option=np.zeros((n+1, n+1))

#For each path
for j in range(0,n+1):
    #For each time step
    for i in range(0,j+1):
        #The nodes are powers of u and d
        Stree[i,j] = S*(u**(j-i))*(d**(i))

        #Depending if is a call or a put
        #we apply a different payoff function
        if otype=="call":
            Intrinsic[i,j]=np.maximum(Stree[i,j]-K,0)
        elif otype=="put":
            Intrinsic[i,j]=np.maximum(K-Stree[i,j],0)
        else:
            print("Wrong option type. Please write call or put")


#For each path
for j in range(n,-1,-1):
    #For each time step
    for i in range(0, j+1):
        if j==n:
            Option[i,j] = Intrinsic[i,j]
        else:
            Option[i,j] = np.exp(-r*dt)*(Option[i,j+1]*p\
                                + Option[i+1,j+1]*(1-p))

            #If it is American, we compare the option at the node
            #with its intrinsic value.
            #Choosing the maximum.
            if EuOrAm=="American":
                Option[i,j]=np.maximum(Intrinsic[i,j],Option[i,j])
```

```
    print("The Option price is", Option [0,0])


    return (u,d,p,Stree, Intrinsic, Option)
```

```
Binomial(5,8,0.02, 0.20,2,2, "put", "American")
```

```
The Option price is 3.0
(1.2214027581601699,
 0.8187307530779818,
 0.5003342283694454,
 array([[5.        , 6.10701379, 7.45912349],
        [0.        , 4.09365377, 5.        ],
        [0.        , 0.        , 3.35160023]]),
 array([[3.        , 1.89298621, 0.54087651],
        [0.        , 3.90634623, 3.        ],
        [0.        , 0.        , 4.64839977]]),
 array([[3.        , 1.89298621, 0.54087651],
        [0.        , 3.90634623, 3.        ],
        [0.        , 0.        , 4.64839977]]))
```

We have completed our first code for valuing options under the Binomial model. The code's purpose is not to be efficient. But to show how useful it is to program. With a few elements, we can code our model. It allows us to increase the accuracy, for example, by increasing the time step number or by automating the calculations.

5.5 CHAPTER'S DIGEST

When valuing simple financial derivatives, such as futures or forwards, we can find the market price using no-arbitrage assumptions. A model is not needed. The price is constructed by combining the asset with some cash. However, this is not possible when we value financial derivatives with an option. In this case, we need a random model. The simplest model is the Binomial model. In this model, today's price can change to two different prices at a future time. Its dynamics are not realistic but will help us understand how option pricing works.

The option price depends on the width of these two prices. We call delta the relationship between the amplitude of payoff prices and asset prices. Using delta, we can find the option value over a period, or time step, using no-arbitrage assumptions. We need a portfolio with delta assets and some cash.

There is another way to value an option. We find an imaginary probability where the expected scenario is the forward value. This probability is called the risk-neutral probability. Although it may seem strange, this is a powerful technique because it allows us to find the price of an option using mathematical expectations.

It is possible to extend the binomial trees to more periods, finding more scenarios. In particular, if we expand the tree to n periods, we will have n+1 scenarios. Once we obtain the future prices, we can calculate the option value using discounted expected values backward.

Options can be European or American. The difference is that the second gives us the right to execute the option at any time before expiration. We can adapt the Binomial pricing scheme by including this feature. For each node, we compare the intrinsic value and the option value, choosing the maximum that refers to the optimal price.

The option price is a function that depends on the spot, the strike, the maturity time, the volatility, and the interest rates. When we change any of these values, the option price changes. These changes are the sensitivity of the price to a variable variation. In mathematical terms, they are function derivatives. In financial terms, we call them the Greeks. These sensitivities will help us explain the price changes in two different periods.

5.6 EXERCISES

1. A stock price is currently at 40 €. In one month, the price will be either 42€ or 38€. The risk-free interest rate is 8% per annum, continuously compounded. We have a call option with a strike price of 39€.

    (a) How many shares we will buy or sell to create our risk-less portfolio? **Solution: 0.75**

    (b) How much cash do we need to replicate the portfolio? **Solution: −28.31**

    (c) How much is it value the one-month European Call option? **Solution: 1.69**

    (d) Which is the value of a put option? **Solution: 0.43**

    (e) Test call-put parity.

2. Calculate the possible values of an asset using a binomial model of 5 steps taking into account that $S_0 = 100$, r = 5%, $\sigma = 10\%$ and the maturity is in one year. Calculate the expected value at the 5th step. Test it with the forward value. **Solution: 105.13**

3. A stock price is currently at 100€. Over the next year it is expected to go up by 10% or down by 10%. The risk free rate is 8% annually, continuously compounded. What is the value of a one-year European call with a strike price of 100€? What is the value of a European put? Verify that put-call parity holds. **Solution: call=8.46, put=0.77.**

4. The Australian dollar is currently worth 0.7 dollars and this exchange rate has a volatility of 5%. The Australian risk-free rate is 8% and the US risk-free rate is 5%. Assuming a one-step binomial model, what is the price of the one-year corresponding European call with strike price $K = 0.7$? (Obs: consider US dollar as the domestic currency). **Solution: call = 0.006\$**

5. Calculate the price of an American put written over a non-divided paying stock, with 6-month maturity and strike price K = 50 in a 2-period binomial tree (= each period of the tree takes 3 months). The price of the underlying stock at $t = 0$ is $S_0 = 50$. During each period, the price of the underlying can go up or down by 10%. Continuously compounded risk free return is 10% annually. **Solution: 1.91**

CHAPTER 6

A Continuous-time Pricing Model

The Binomial model is an excellent pricing model, although the scenarios considered are not good enough to reproduce real market data. This drawback allows us to create a simple and visual pricing scheme that will allow us to play with different ideas, such as how to adapt the scheme to value an American option.

But, sometimes, we need a more elaborate model that can describe how the asset moves over time, a continuous-time model. Here we have the Black-Scholes-Merton model. It is not only the most classic model but also the pillar of almost any more advanced model.

6.1 CREATING SOME INTUITION

We are interested in how the assets move over time. Instead of looking at the price evolution, we will look at its counterpart, the price returns. Note that price returns are comparable across assets, while price evolution is not. We will denote the asset price on the ith day by P_i, then the return from day i to day $i+1$ is

$$R_i = \frac{P_{i+1} - P_i}{P_i}.$$

The accumulate return from day 0 and the ith day is

$$\frac{P_i - P_0}{P_0} = (1 + R_0) \cdot \ldots \cdot (1 + R_{i-1}) - 1 = \prod_{j=0}^{i-1}(1 + R_j) - 1.$$

DOI: 10.1201/9781003266730-6

Now, we can consider the log-returns

$$R_i = \ln\left(\frac{P_{i+1}}{P_i}\right).$$

In this case, the accumulate returns from day 0 and the ith day is

$$\ln\left(\frac{P_{i-1}}{P_0}\right) = \ln\left(\frac{P_1}{P_0} \cdot \frac{P_2}{P_1} \cdots \cdots \frac{P_{i-1}}{P_{i-2}}\right) = \ln\left(\frac{P_1}{P_0}\right) + \ldots + \ln\left(\frac{P_{i-1}}{P_{i-2}}\right)$$

Watch this: to accumulate returns, we have to use multiplications, but thanks to the properties of the logarithm, we can accumulate with additions. This is one of the advantages of using logarithms and exponentials.

We plot the historical price of a fictitious asset. The price has been constructed from the returns of different real stocks using a historical bootstrapping technique. Thus, we continue to observe market characteristics.

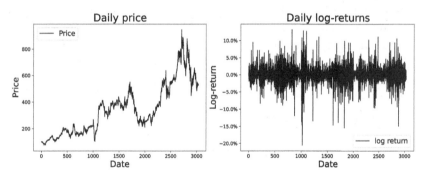

Figure 6.1 Price and log-return of a fictitious asset.

The Figure 6.1 is very instructive. On the left side, we can see the price's evolution. On the right hand side, we have the log-returns. On average, they are close to zero and move erratically from positive to negative with similar intensity, except for a few periods when the intensity changes.

Focusing on log-returns, the above features give us insight into the underlying model:

- The model is random as we go up and down seemingly without any sense.

- On average it is close to zero.
- By accumulating days, we obtain a linear trend.

To enrich the idea of how the log-returns behave, we will make the histogram (see Figure 6.2).

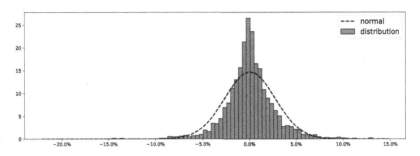

Figure 6.2 Histogram with Normal Distribution Curve.

We can see that the empirical distribution is symmetric with the mean above zero, even though the left tail of the distribution is longer than the right tail. We adjust the empirical distribution with a Normal distribution. Visually, the normal distribution seems to fit well, except for the distribution center. Beyond the fitting properties, this distribution has rich mathematical properties that will be useful.

Therefore, it is logical to assume that the returns behave like this (see Figure 6.3). Furthermore, for the sake of simplicity, it is common to assume that each day the asset log-returns will have the same distribution regardless of what has happened in the past. That is, the returns have no memory. When a random process exhibits this behavior, we call it an independent and identically distributed random variable, also known as an *iid*.

Recall a property of normal distribution. If X_1 and X_2 are two independent normal random variables, with means μ_1, μ_2 and standard deviations σ_1, σ_2, then their sum $X_1 + X_2$ will also be normally distributed, $\mu_1 + \mu_2$ and variance $\sigma_1^2 + \sigma_2^2$. In other words,

$$X_1 + X_2 \sim \mathcal{N}\left(\mu_1, \sigma_1^2\right) + \mathcal{N}\left(\mu_2, \sigma_2^2\right) \sim \mathcal{N}\left(\mu_1 + \mu_2, \sigma_1^2 + \sigma_2^2\right)$$

112 ■ A Continuous-time Pricing Model

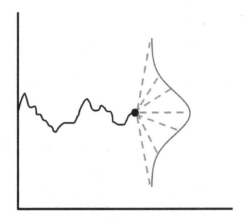

Figure 6.3 Model intuition.

In our case, each day the log-returns follow a normal distribution with the same average and the same standard deviation which is the equivalent of having sums of normal distribution, so

$$\mathcal{N}\left(\mu,\sigma^2\right) + \underbrace{\ldots}_{\text{n days}} + \mathcal{N}\left(\mu,\sigma^2\right) \sim \mathcal{N}\left(n\mu, n\sigma^2\right)$$

After n-days, we have a normal distribution with mean $n\mu$ and standard deviation $\sqrt{n}\sigma$. The trend is accumulated linearly and the volatility is by the inverse quadratic growth. Note that we just found a rule for re-scale the parameters to different horizons (see Figure 6.4).

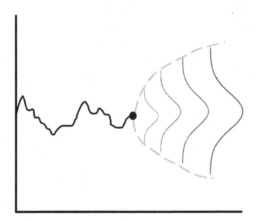

Figure 6.4 Accumulating days intuition.

Remember that under a risk-neutral measure, the prices growth must match the forward value, therefore μ must be the risk-free rate (when the asset does not pay dividends).

All the above reasoning leads us to consider a model somehow connected to the normal distribution of the returns. Nevertheless, we recall that the primary goal in derivative pricing is not only to find a model that reproduces prices but rather to price assets and standard options (which are the inputs in our model calibration) in the sense that arbitrage opportunities are not allowed. Then, the calibrated model will be used to price exotics. This is not a trivial problem, and several solutions have been proposed in the literature, even when nowadays most models are based on the Black-Scholes-Merton framework. Let us see in the next section the different approaches proposed in the literature, from the early XIX century, and why the Black-Scholes-Merton has become the basic model in option pricing.

6.2 THE BLACK-SCHOLES-MERTON FRAMEWORK

Although financial contracts evolved, it was not until 1900 that the history of mathematical modeling of financial markets began. It started with a French mathematician, Louis Bachelier. The first quant. Bachelier presented the first model for price options in his Ph.D. thesis 'Théorie de la spéculation' in 1900, (Bachelier, 1900). In this work, Bachelier simulates the asset's evolution through a process that looks like a random walk, in fact, that is the name of the process. One of the pitfalls of the model is that it produces negative prices. Unfortunately, Bachelier's work was not appreciated at his time and remained unknown for decades. A few decades later, in the 50s, pricing warrants and options become an exciting topic to work on again. Independently of Bachelier's work, the astronomer M.F.M. Osborne (Osborne, 1959), made a similar model approach, but this time for the log-returns. Also, statistician Jimmy Savage retrieved Bachelier's work and sent it to different friends. Fortunately, one of those postcards found its way to Paul Samuelson. Inspired by Bachelier's work, he extended it in (Samuelson, 1965). At that time, several formulas appeared with different approaches and parameters for example (Sprenkel, 1961) or (Kassoui and Thorp, 1967). Despite that, it was not until 1973, the year the world's first listed options exchange opened in Chicago, the Chicago Board Options Exchange (CBOE), the famous Black-Scholes-Merton model was published using no-arbitrage assumptions, see (Black and Scholes, 1973) and (Merton, 1973).

In this framework, we will assume that:

- There are no arbitrage opportunities.

- We can lend or borrow any amount of cash at the risk-free rate.

- We can buy or sell any amount of the stock.

- There are not transactions costs or fees.

One of the main advantages of the Black-Scholes-Merton model is that it does not allow, under some conditions, for arbitrage opportunities, as we discuss in the following section.

6.3 THE BLACK-SCHOLES-MERTON EQUATION

We have seen some intuition about the normality of asset returns. Let us see how a non-arbitrage replication argument also leads us, under some conditions, to this model. Imagine you are tasked with pricing and risk managing an option, for example, a European Call option. We will refer to the price as $P(t, S_t)$.

We can replicate this option by using a replication portfolio based on:

- Delta asset units.

- A risk-free investment of $P(t, S_t) - \Delta S_t$.

If the replication portfolio works, the option and the replication portfolio must move equally. So, if we are long in option and short on the replication portfolio the profit is zero. Numerically, we will write

$$\underbrace{P(t, S_t)}_{Derivative} = \underbrace{\Delta S_t}_{Stocks} + \underbrace{(P(t, S_t) - \Delta S_t)}_{Cash}.$$

After one period, we know that

- The asset market moves. If the asset changes to $S_t + \delta S_t$, then, the return is:

$$\underbrace{\Delta(S_t + \delta S_t)}_{New Period} - \underbrace{\Delta S_t}_{Past Period} = \Delta \delta S_t.$$

- We have invested on a bank account an amount of $(P(t, S_t) - \Delta S_t)$. The return is the interest rates over the amount deposited during the period δt:

$$\underbrace{e^{r\delta t}\left(P(t, S_t) - \Delta S_t\right)}_{New Period} - \underbrace{\left(P(t, S_t) - \Delta S_t\right)}_{Past Period} \approx r\delta t \left(P(t, S_t) - \Delta S_t\right).$$

To obtain this approximation, we use the following Taylor series approximation:

$$e^{r\delta t} \approx (1 + r\delta t).$$

- The derivative changes

$$\underbrace{P(t + \delta t, S_t + \delta S_t)}_{New Period} - \underbrace{P(t, S_t)}_{Past Period}.$$

Writing all the movements together

$$P(t + \delta t, S_t + \delta S_t) - P(t, S_t) = \Delta \delta S_t + r\delta t \left(P(t, S_t) - \Delta S_t\right). \quad (6.1)$$

We can approximate the derivative movement by a Taylor series approximation

$$\begin{aligned} P(t + \delta t, S_t + \delta S_t) - P(t, S_t) &= \partial_t P(t, S_t)\delta t + \partial_S P(t, S_t)\delta S_t \\ &+ \frac{1}{2}\partial_S^2 P(t, S_t)(\delta S_t)^2. \end{aligned}$$

Remember that $\Delta = \partial_S P(t, S_t)$, then

$$P(t + \delta t, S_t + \delta S_t) - P(t, S_t) = \partial_t P(t, S_t)\delta t + \Delta \delta S_t + \frac{1}{2}\partial_S^2 P(t, S_t)(\delta S_t)^2.$$

If we assume that the variance of the price is

$$\left(\frac{\delta S_t}{S_t}\right)^2 \approx \sigma^2 \delta t,$$

or, equivalently

$$(\delta S_t)^2 \approx S^2 \sigma^2 \delta t.$$

Then, substituting the derivative movement approximation into (6.1), we have

$$\partial_t P(t, S_t)\delta t + \frac{1}{2}\partial_S^2 P(t, S_t)S_t^2\sigma^2\delta t = r\delta t\left(P(t, S_t) - \Delta S_t\right).$$

Re-arrenging and applying again the change $\Delta = \partial_S P(t, S_t)$ we obtain

$$0 = \partial_t P(t, S_t)\delta t + r\partial_S P(t, S_t)S_t\delta t + \frac{1}{2}\partial_S^2 P(t, S_t)S_t^2\sigma^2\delta t - rP(t, S_t)\delta t.$$

This is the Black-Scholes-Merton equation. This equation imposes a relationship between the partial derivatives of a function. These types of equations are Partial Differential Equations, or PDEs. This equation is also known as the Black-Scholes-Merton PDE. The solution of this PDE is the so-called Black-Scholes-Merton formula that we discuss in the following section. Notice that we have arrived at this equation just from non-arbitrage arguments and assuming that

$$\left(\frac{\delta S_t}{S_t}\right)^2 \approx \sigma^2\delta t.$$

Even when this latest hypothesis is not realistic and it is weakened in several models used in practice, the above arguments justify that option prices are based on the Black-Scholes-Merton PDE.

Although it can seem strange depending on your background, physicists and engineers use this type of equation extensively to model different phenomena such as heat, sound, and fluid dynamics, ... That explains why in the mid-1970s and 1980s many physicists switched from physicists to financial engineers.

6.4 THE BLACK-SCHOLES-MERTON FORMULA

The Black-Scholes-Merton formula, deduced as the solution of the Black-Scholes-Merton equation, is an analytical formula that describes parsimoniously market option prices. The main drivers are the ratio between the stock price and the strike, also known as moneyness; the level of interest rates; and the constant volatility. Today it is the most popular formula, becoming the benchmark model. This is not only for its simplicity, but also for the techniques used to derive the formula.

The formula for a European Call option is

$$Call(S_0, K, t, T, r, \sigma) = S_0\mathcal{N}(d_1) - Ke^{-r(T-t)}\mathcal{N}(d_2), \qquad (6.2)$$

where $\mathcal{N}$ denotes the normal distribution function and

$$d_1 = \frac{\ln\left(\frac{S_0}{K}\right) + r(T-t)}{\sigma\sqrt{T-t}} + \frac{\sigma\sqrt{T-t}}{2}, \qquad (6.3)$$

$$d_2 = \frac{\ln\left(\frac{S_0}{K}\right) + r(T-t)}{\sigma\sqrt{T-t}} - \frac{\sigma\sqrt{T-t}}{2}. \qquad (6.4)$$

Note that

$$d_2 = d_1 - \sigma\sqrt{T-t}. \qquad (6.5)$$

In a similar way, the formula for a European Put option is

$$Put(S_0, K, t, T, r, \sigma) = Ke^{-rT}\mathcal{N}(-d_2) - S_0\mathcal{N}(-d_1). \qquad (6.6)$$

Example 6.1

Price a European Call option with $S_0 = 100$, $K = 105$, $r = 0.02$, $\sigma = 0.1$, and $T = 1$.

$$\begin{aligned}
\text{Price} &= 100\mathcal{N}\left(\frac{\ln\left(\frac{100}{105}\right) + 0.02}{0.1} + \frac{0.1}{2}\right) \\
&\quad - 105e^{-0.02}\mathcal{N}\left(\frac{\ln\left(\frac{100}{105}\right) + 0.02}{0.1} - \frac{0.1}{2}\right) \\
&= 2.75.
\end{aligned}$$

Nowadays, this formula is very simple to calculate. The only difficulty is the Normal distribution calculation which is available in almost all computer programs. We are going to see how to do it in Python.

```
#We load the Packages needed
import scipy.stats as si
import numpy as np

#We define the Black-Scholes-Merton formula
def BS(S, K, T, r, sigma, option):

    #S: spot price
    #K: strike price
    #T: time to maturity
    #r: interest rate
    #sigma: volatility of underlying asset
```

```
    d1 = (np.log(S / K) + (r + 0.5 * sigma ** 2) * T)\
        / (sigma * np.sqrt(T))
    d2 = (np.log(S / K) + (r - 0.5 * sigma ** 2) * T)\
        / (sigma * np.sqrt(T))

    if option == 'call':
        result = (S * si.norm.cdf(d1, 0.0, 1.0) \
                 - K * np.exp(-r * T) * si.norm.cdf(d2, 0.0, 1.0))
    if option == 'put':
        result = (K * np.exp(-r * T) * si.norm.cdf(-d2, 0.0, 1.0)\
                 - S * si.norm.cdf(-d1, 0.0, 1.0))

    return result
```

We test our code with the same option we value by hand.

```
BS(100, 105, 0.02, 0.1, 1)
```

```
2.7519486233365313
```

Now, we can play around. We are going to see how the price of an option changes when the maturity increases. We observe the price is closer to the payoff for short-maturities (see Figure 6.5).

```
#We load the Packages needed
import numpy as np
import scipy.stats as si
import matplotlib.pyplot as plt

#We define the plot size
fig = plt.figure(figsize=(15,5))

#We create a vector with spot values from 75 to 125
x=np.linspace(75,125)

#We plot all the moneyness with maturity 1/12 years
plt.plot(x, BS(x, 100, 1/12, 0.05, 0.20, 'call'), 'k', label='1 Month')
#We plot all the moneyness with maturity 1 year
plt.plot(x, BS(x, 100, 1, 0.05, 0.20, 'call'), 'k--', label='1 Year')
#We plot all the moneyness with maturity 2 years
plt.plot(x, BS(x, 100, 2, 0.05, 0.20, 'call'), 'k.', label='2 Years')
#We plot all the moneyness with maturity 5 years
plt.plot(x, BS(x, 100, 5, 0.05, 0.20, 'call'), 'k_', label='5 Years')
#We write the legend
plt.legend()
#We show the plot
plt.show()
```

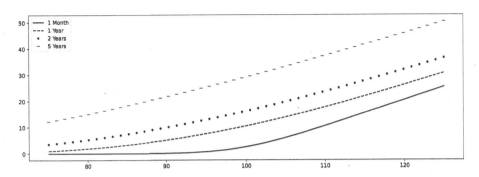

Figure 6.5 Black-Scholes-Merton formula for different maturities.

We can change the way we look at the prices. We can see it from a 3D perspective (see Figure 6.6).

```
#We load the Packages needed
import matplotlib.pyplot as plt
import numpy as np

#Define the plot size
fig = plt.figure(figsize=(25,15))

# Make data

#Create a vector with spot values from 75 to 125
X = np.arange(75, 125, 1)

#Create a vector with time values from 0.25 to 10
Y = np.arange(0.25, 5, 0.25)

#Combine
X, Y = np.meshgrid(X, Y)

#Evaluate the BS formula in these cases
Z = BS(X, 100, Y, 0.05, 0.20, 'call')

#Prepare the 3D plot
ax = plt.axes(projection='3d')
surf=ax.plot_wireframe(X, Y, Z, color='black')

# Set axes label
ax.set_xlabel('Spot', fontsize=18)
ax.set_ylabel('Time', fontsize=18)
ax.set_zlabel('Price', fontsize=18)

#Size axis numbers
ax.tick_params(axis='both',length=5,width=2,labelsize=14)

plt.show()
```

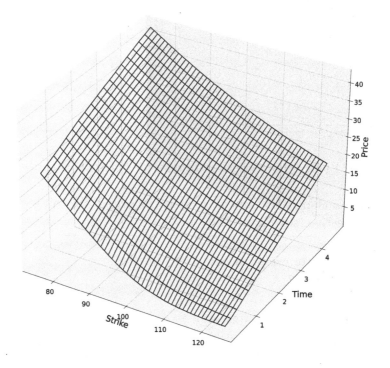

Figure 6.6 Black-Scholes-Merton formula in 3D.

6.5 THE BLACK-SCHOLES-MERTON MODEL FROM A PROBABILISTIC PERSPECTIVE

In the first section, we have seen how to model prices by observing how returns behave. Later, we find analytical insight deriving an option and its portfolio replication. Now it is time to see the Black-Scholes-Merton formula from a probabilistic point of view. All these perspectives are connected.

We introduce the Brownian Motion. This motion represents the random motion of particles suspended in a medium such as a liquid or a gas. In 1827, it was observed by the Scottish botanist Robert Brown, who looked through a microscope at pollen grains suspended in water. In 1905, Albert Einstein modeled it. In mathematics, it is known as the Wiener process, W_t, a continuous-time stochastic process. It has the following properties:

- The process depends on the time variable that is the reason of the subscript t.

- At time zero, $W_0 = 0$.

- The sample trajectories $t \mapsto W_t$ are continuous (with probability 1).

- For any finite sequence of times $t_0 < t_1 < \cdots < t_n$, the increments

$$W_{t_1} - W_{t_0}, W_{t_2} - W_{t_1}, \cdots, W_{t_n} - W_{t_{n-1}}$$

are independent.

- It has stationary increments. For any $t_1 < t_2$, $W_{t_2} - W_{t_1} \sim W_{t_2-t_1} - W_0$.

- W has Gaussian increments. $W_t - W_0 \sim \mathcal{N}(0, t)$.

Some of the properties are connected with what we have seen statistically. For example,

- Each period of returns is independent of the previous one.

- All the periods, follow the same distribution.

- The log-returns follow a normal distribution.

Another important feature of this mathematical object is that it is continuous, but not differentiable. Not in the classical sense. Instead of being a smooth function, it is noisy. This feature will make the process look very similar to reality.

The asset prices are modeled by the following dynamics:

$$dS_t = \underbrace{r \; S \; dt}_{trend} + \underbrace{\sigma \; S \; dW_t}_{amplitude \; scenarios}. \tag{6.7}$$

This equation is a Stochastic Differential Equation, SDE, and its dynamics corresponds to what we observe in Figure 6.1. To deal with these kinds of equations, we need to use the theory developed by Stochastic Calculus. One of the fundamental theorems is the Itô formula, the stochastic calculus counterpart of the 'chain rule'.

122 ■ A Continuous-time Pricing Model

Using the Itô formula to (6.7), we obtain

$$d\ln S_t = \left(r - \frac{1}{2}\sigma^2\right)dt + \sigma dW_t. \tag{6.8}$$

This equation means that a 'small variation' of the price has a trend of rt with an amplitude led by a normal distribution with variance $\sigma^2 t$. It is equivalent to writing that

$$S_t = S_0 \exp\left(\left(r - \frac{1}{2}\sigma^2\right)t + \sigma dW_t\right). \tag{6.9}$$

We can re-express the above formula as

$$S_t = S_s \exp\left(\left(r - \frac{1}{2}\sigma^2\right)(t-s) + \sigma dW_{t-s}\right).$$

where $s < t$, but without loss of generality, we will assume $s = 0$.

That is, the asset prices are exponentials of normal variables. In this case, we say that asset prices are **log normal**. These formulas are very useful. On the one hand, we can do some calculus using basic notions of probability. On the other hand, we can simulate the process using numerical methods.

We can calculate the probability of the option being exercised that will happen when

$$S_T > K \;=\; S_0 \exp\left(\left(r - \frac{1}{2}\sigma^2\right)T + \sigma dW_T\right) > K$$

Remeber that $W_0 = 0$, therefore, we have that

$$W_T = W_T - W_0 \sim \sqrt{T}\mathcal{N}(0,1) = \sqrt{T}z$$

with $z \sim \mathcal{N}(0,1)$. Using this on the previous formula, we have that

$$\begin{aligned}
S_T > K \;&=\; S_0 \exp\left(\left(r - \frac{1}{2}\sigma^2\right)T + \sigma\sqrt{T}z\right) > K \\
&= \exp\left(\left(r - \frac{1}{2}\sigma^2\right)T + \sigma\sqrt{T}z\right) > \frac{K}{S_0} \\
&= \left(r - \frac{1}{2}\sigma^2\right)T + \sigma\sqrt{T}z > \ln\left(\frac{K}{S_0}\right) \\
&= z > \frac{\ln\left(\frac{K}{S_0}\right) - \left(r - \frac{1}{2}\sigma^2\right)T}{\sigma\sqrt{T}}.
\end{aligned} \tag{6.10}$$

Therefore,

$$\mathbb{P}(S_T > K) = \mathbb{P}\left(z > \frac{\ln\left(\frac{K}{S_0}\right) - \left(r - \frac{1}{2}\sigma^2\right)T}{\sigma\sqrt{T}}\right). \quad (6.11)$$

By symmetry of the normal distribution, we have that

$$\mathbb{P}(z > -x) = \mathbb{P}(z < x) = \mathcal{N}(x).$$

Using it above and the following fact

$$-\ln\left(\frac{K}{S_0}\right) = -(\ln K - \ln S_0) = \ln S_0 - \ln K = \ln\left(\frac{S_0}{K}\right),$$

we obtain that

$$\begin{aligned}\mathbb{P}(S_T > K) &= \mathbb{P}\left(z < \frac{\ln\left(\frac{S_0}{K}\right) + \left(r - \frac{1}{2}\sigma^2\right)T}{\sigma\sqrt{T}}\right) \\ &= \mathbb{P}(z < d_2) \\ &= \mathcal{N}(d_2).\end{aligned}$$

Remember that a probability distribution is the mathematical function that gives the probabilities of occurrence of different possible outcomes for an experiment or event. When the probability distribution is absolutely continuous, then the probability of a random process X belonging to $[a, b]$ is given by the integral

$$\mathbb{P}(a \leq X \leq b) = \int_a^b f(x)dx$$

where f is the density (or probability density function) a non-negative integral function (see Figure 6.7).

Graphically, the density function is the contour and the probability is the area covered between two points.

In the normal distribution case, the density is

$$f(x) = \frac{1}{\sigma\sqrt{2\pi}} \exp\left(-\frac{1}{2}\left(\frac{x-\mu}{\sigma}\right)^2\right) \quad (6.12)$$

where μ is the mean and σ is the standard deviation.

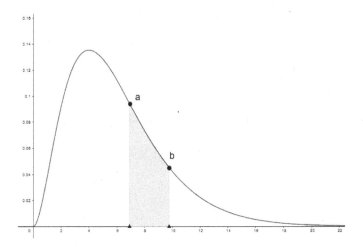

Figure 6.7 Probability Distribution and the density function.

In the discrete case, the mathematical expectation is a weighted average. In the continuous case, given a continuous random variable X with density $f(x)$, the mathematical expectation of $g(x)$ is

$$\mathbb{E}\left[g(x)\right] = \int_{-\infty}^{\infty} g(x)f(x)dx.$$

So, we are changing the contour by the value $g(x)$.

We can write the call option in terms of mathematical expectations,

$$\exp\left(-rT\right) \mathbb{E}\left[(S_T - K, 0)_+\right] \tag{6.13}$$

where $(S_T - K, 0)_+$ is the positive part of the payoff.

Using some probability properties, we shown how to obtain the Black-Scholes-Merton formula.

$$\exp\left(-rT\right) \mathbb{E}\left[(S_T - K, 0)_+\right] = \exp\left(-rT\right) \int_{-\infty}^{\infty} (S_T - K, 0)_+ \phi(z)dz.$$

where $\phi(z)$ is the normal density function with mean 0 and standard deviation 1.

The Black-Scholes-Merton Model ■ 125

The integral is positive when (6.10) holds, then

$$\exp(-rT)\,\mathbb{E}\left[(S_T-K,0)_+\right] = \exp(-rT)\int_{-\infty}^{\infty}(S_T-K,0)_+\,\phi(z)dz$$
$$= \exp(-rT)\int_{-d_2}^{\infty}(S_T-K)\,\phi(z)dz.$$

Note that this integral can be separated in two integrals. On the one hand, we have

$$\exp(-rT)\int_{-d_2}^{\infty}K\phi(z)dz = K\exp(-rT)\int_{-d_2}^{\infty}\phi(z)dz$$
$$= K\exp(-rT)(1-\mathcal{N}(-d_2))$$
$$= K\exp(-rT)\,\mathcal{N}(d_2). \qquad (6.14)$$

In the other hand,

$$\exp(-rT)\int_{-d_2}^{\infty}S_T\phi(z)dz$$
$$= \exp(-rT)\int_{-d_2}^{\infty}S_0\exp\left(\left(r-\frac{1}{2}\sigma^2\right)T+\sigma\sqrt{T}z\right)\phi(z)dz$$
$$= S_0\int_{-d_2}^{\infty}\exp\left(-\frac{1}{2}\sigma^2 T+\sigma\sqrt{T}z\right)\phi(z)dz.$$

Using the normal density function (6.12), we have

$$\exp(-rT)\int_{-d_2}^{\infty}S_T\phi(z)dz$$
$$= \frac{S_0}{\sqrt{2\pi}}\int_{-d_2}^{\infty}\exp\left(-\frac{1}{2}\sigma^2 T+\sigma\sqrt{T}z\right)\exp\left(-\frac{1}{2}z^2\right)dz$$
$$\frac{S_0}{\sqrt{2\pi}}\int_{-d_2}^{\infty}\exp\left(-\frac{1}{2}\sigma^2 T+\sigma\sqrt{T}z-\frac{1}{2}z^2\right)dz. \qquad (6.15)$$

Now, we have to apply the trick 'complete the square' that consist in express the terms inside the exponential as an square

$$-\frac{1}{2}\sigma^2 T+\sigma\sqrt{T}z-\frac{1}{2}z^2 = -\frac{1}{2}\left(\sigma^2 T-2\sigma\sqrt{T}z+z^2\right)$$
$$= -\frac{1}{2}\left(z-\sigma\sqrt{T}\right)^2. \qquad (6.16)$$

Substitying (6.16) in (6.15), we have

$$\exp(-rT)\int_{-d_2}^{\infty}S_T\phi(z)dz = \frac{S_0}{\sqrt{2\pi}}\int_{-d_2}^{\infty}\exp\left(-\frac{1}{2}\exp\left(\left(z-\sigma\sqrt{T}\right)^2\right)\right)dz.$$

Then,

$$\begin{aligned}\exp(-rT)\int_{-d_2}^{\infty} S_T \phi(z)dz &= S_0 \mathbb{P}\left(z - \sigma\sqrt{T} \le -d_2\right) \\ &= S_0 \mathbb{P}\left(z - \sigma\sqrt{T} \ge d_2\right) \\ &= S_0 \mathcal{N}\left(d_2 + \sigma\sqrt{T}\right) \\ &= S_0 \mathcal{N}(d_1). \end{aligned} \quad (6.17)$$

Now, if we use (6.17) and (6.14), we find that the price of (6.13) is the Black-Scholes-Merton formula with $t = 0$.

If you are wondering if the Black-Scholes-Merton equation where we obtained a PDE and the probabilistic view are connected, the short answer is yes, they are. The Feynman-Kac Theorem links the PDE solution with the conditional expectation of a random process following an SDE. But this is out of the scope of the book.

6.6 THE BLACK-SCHOLES-MERTON PRICE AND THE BINOMIAL PRICE

Now, we have more than a way to calculate a vanilla European call or put option. On the one hand, we have the Binomial model. On the other hand, the Black-Scholes-Merton formula. There is a relationship between both formulas. When the time step tends to zero, i.e. we have infinite number of steps, the Binomial model converges to the Black-Scholes-Merton price.

We can see it visually. For example, given a European call option with $S_0 = 100, K = 100, r = 0, \sigma = 0.25, T = 1$, we will value the Binomial model from 2 steps to 100 (see Figure 6.8).

```
#We create a value to save all the Binomial model values
values=np.zeros((99))

#We define the initial values
S=100
K=100
r=0
sigma=0.25
T=1
otype="call"
```

```
#With a for, we evaluate the Binomial model
# with different steps. In particular, from
# 2 to 100
for i in range(2,101):
    values[i-2]=EuBinomial(S,K,r,sigma,T,i,otype)

#We create a figure
fig = plt.figure(figsize=(15,5))
#We plot all the values of the Binomial model
plt.plot(np.arange(2, 101, 1), values, 'k', label='Binomial Call')
#We plot a constant line with the value of the BS price
plt.hlines(y=BS(100, 100, 1, 0.00, 0.25, 'call'),xmin=2, xmax=100,\
 linewidth=3,linestyles=':', color='k', label='BSM Call')
#We write the legend
plt.legend()
#We show the plot
plt.show()
```

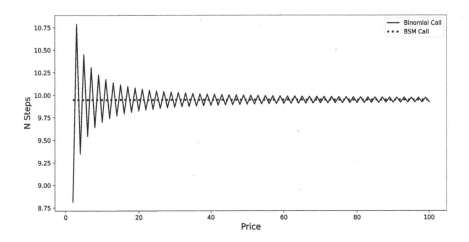

Figure 6.8 ATM Delta of a call as a function of T.

In this case, the Black-Scholes-Merton price is 9.9476.

6.7 THE GREEKS IN THE BLACK-SCHOLES-MERTON MODEL

As we have seen in Chapter 5, the Greeks explain how much the price function changes when a variable, or 'input', changes. In the case of Binomial, to calculate it, in most cases, we use the tree nodes. Now, we are in a continuous model, so we can calculate its derivatives as we have done for other functions in an Analysis course. In this section, we will assume that $t = 0$, in other words, we are going to look at the behavior from today's perspective.

6.7.1 Delta

The Delta (Δ) is the derivative of the Black-Scholes-Merton price with respect to the asset price.

The call option delta is given by

$$\Delta Call(S_0, K, 0, T, r, \sigma) = \mathcal{N}(d_1), \tag{6.18}$$

and the put option delta by

$$\Delta Put(S_0, K, 0, T, r, \sigma) = N(d_1) - 1, \tag{6.19}$$

where d_1 is given by 6.3.

The computation of Delta can be easily implemented in Python. For example, in the case of a call a simple script could be as follows:

```
#We load the packages we are going to use
import numpy as np
import scipy.stats as ss

#We define d1 and d2 as separate functions.
#The idea is to use it for the BSM as well as the Greeks

def d1(S0,K,r,sigma,T):
    #Function to calculate d1
    #S0: Spot price
    #K: Strike price
    #r: risk-free rate
    #sigma: volatility
    #T: Time to maturity
    return (np.log(S0/K)+(r+sigma**2/2)*T)/(sigma*np.sqrt(T))

def d2(S0,K,r,sigma,T):
    #Function to calculate d2
    #S0: Spot price
    #K: Strike price
    #r: risk-free rate
    #sigma: volatility
    #T: Time to maturity
    return(np.log(S0/K)+(r-sigma**2/2)*T)/(sigma*np.sqrt(T))

def DeltaCall(S0,K,r,sigma,T):
    #Function to calcula the call option delta
    #S0: Spot price
```

```
#K: Strike price
#r: risk-free rate
#sigma: volatility
#T: Time to maturity
return ss.norm.cdf(d1(S0,K,r,sigma,T))
```

The good thing about having a mathematical expression is that we can imply different behaviors of the call option delta.

For example, when a call option is ATM by definition $\ln(S_t) - \ln(K) = 0$, then

$$d_1 = \frac{r\sqrt{T}}{\sigma} + \frac{\sigma\sqrt{T}}{2}. \qquad (6.20)$$

If the option expires soon, the time to T will be close to zero. If we look at the limit as $T \to 0$, we can see that d_1 approaches zero and is an increasing function of T. Then, for ATM options, the call option delta tends to $\frac{1}{2}$ for short maturities, and it is an increasing function of T, as we can see in the following plot, corresponding to the call option delta with $S_0 = K = 100$, $r = 0$ and $\sigma = 0.2$ (see Figure 6.9).

```
#We create a figure
fig = plt.figure(figsize=(12,6))

#We plot the Call option delta for a maturity grid
#between 0.01 and 10years
plt.plot(np.linspace(0.001,10), \
        DeltaCall(100,100,0,0.2,np.linspace(0.001,10)), 'k')

#We write a label on the x-axis
plt.xlabel('Delta', fontsize=14)

#We write a label on the y-axis
plt.ylabel('Time (years)', fontsize=14)

#We show the plot
plt.show()
```

130 ◼ A Continuous-time Pricing Model

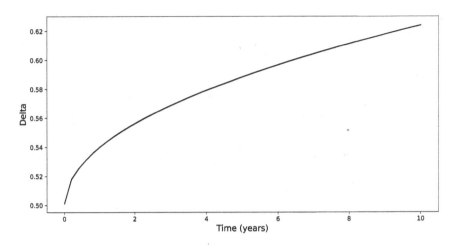

Figure 6.9 ATM Delta of a call as a function of T.

But what happens when the option is ITM, $S_0 > K$. Then, $d_1 \to \infty$ as $T \to 0$, which implies that the short-term call option delta is near 1 (see Figure 6.10).

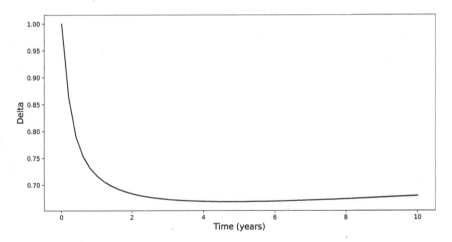

Figure 6.10 In-the-money Delta ($S_0 = 110$) of a call as a function of T.

Similarly, if the option is OTM, $S_0 < K$, $d_1 \to -\infty$ as $T \to 0$, which implies that the short-term call option delta is near 0, as we can see in the following plot with the same parameters as before (see Figure 6.11).

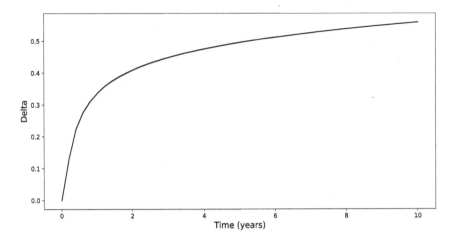

Figure 6.11 Out-of-the-money Delta ($S_0 = 90$) of a call as a function of T.

We can see how the call option delta changes depending on the strike and time doing a 3D plot (see Figure 6.12).

```
#We load the Packages needed
import matplotlib.pyplot as plt
import numpy as np

#Define the plot size
fig = plt.figure(figsize=(20,15))

# Make data

#Create a vector with strike values from 70 to 130
X = np.linspace(70, 130)

#Create a vector with time values from 0.001 to 10
Y = np.linspace(0.001, 10)

#Combine
X, Y = np.meshgrid(X, Y)

#Evaluate the BS formula in these cases
Z = DeltaCall(100,X,0,0.2,Y)

ax = plt.axes(projection='3d')
surf=ax.plot_wireframe(X, Y, Z, color='black')

# Set axes label
ax.set_xlabel('Strike', fontsize=18)
ax.set_ylabel('Time', fontsize=18)
```

```
ax.set_zlabel('Delta', fontsize=18)

#Size axis numbers
ax.tick_params(axis='both',length=5,width=2,labelsize=14)

plt.show()
```

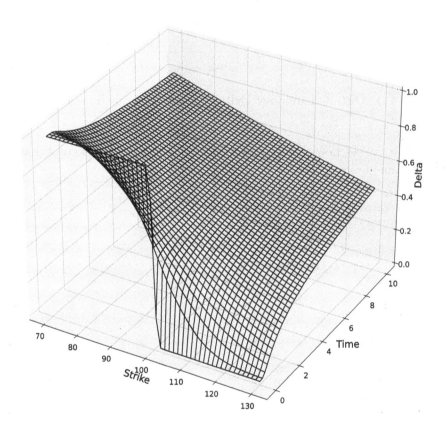

Figure 6.12 Call option delta surface.

6.7.2 Theta

Theta, Θ, is the derivative of the option price with respect to the passage of time.

The call option theta is given by

$$\Theta Call(S_0, K, 0, T, r, \sigma) = -\frac{S_0 \mathcal{N}'(d_1)\sigma}{2\sqrt{T}} - rKe^{-rT}\mathcal{N}(d_2), \quad (6.21)$$

and the put option theta by

$$\Theta Put(S_0, K, 0, T, r, \sigma) = -\frac{S_0 \mathcal{N}'(d_1)\sigma}{2\sqrt{T}} + rKe^{-rT}\mathcal{N}(-d_2), \quad (6.22)$$

where d_2 is given by (6.4).

A Python script for the Θ of a put is given by

```
def ThetaPut(S0, K, e, sigma, T):
    #Function to calculate Put option theta
    #S0: Spot price
    #K: Strike price
    #r: risk-free rate
    #sigma: volatility
    #T: Time to maturity

    #We calculate the probality density func.
    pdf=ss.norm.pdf(d1(S0,K,r,sigma,T))

    #We calculate the cummulative density func.
    cdf=ss.norm.cdf(-d2(S0,K,r,sigma,T))

    return -S0*sigma/(2*np.sqrt(T))*pdf+ r*K*np.exp(-r*T)*cdf
```

In the case of a call option Θ is always negative, unless the risk-free rate is negative. This means that the option decreases when we are near maturity. Equivalently, the option price is higher for longer maturities.

A more interesting case is the put option theta. In this case, if $r \neq 0$, the Θ is not always negative. To be negative we need

$$-\frac{S_0 \mathcal{N}'(d_1)\sigma}{2\sqrt{T}} + rKe^{-rT}\mathcal{N}(-d_2) < 0.$$

That is,

$$rKe^{-rT}\mathcal{N}(-d_2) < \frac{S_0 \mathcal{N}'(d_1)\sigma}{2\sqrt{T}}$$

or, equivalently,

$$\frac{S_0}{K} > \frac{2r\sqrt{T}e^{-rT}\mathcal{N}(-d_2)}{\mathcal{N}'(d_1)\sigma}.$$

That is, the ratio $\frac{S_0}{K}$ has to be bigger than some positive constant that depends on r and on the time to maturity. We can see this effect in Figure 6.13 with the parameters $K = 100$, $r = 0.05$, $\sigma = 0.2$, $T = 1$

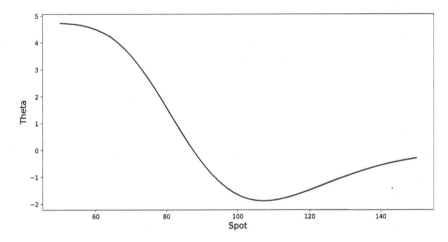

Figure 6.13 Theta for a put as a function of S_0.

Note that Θ is positive for very ITM put options. This is because you are closer to getting the 'intrinsic option value' and it is more difficult for volatility goes against you.

We can plot a theta surface depending on the changes in the spot and the time to maturity. We can observe that put options have a higher theta close to the expiration, especially ATM, but it becomes flat for long maturities (see Figure 6.14).

6.7.3 Gamma

Gamma, Γ, is the second derivative of the option price with respect to S_0. That is, it is the derivative of the Delta with respect to S_0. We will see later that Γ and ϑ are closely related, and they are important in the study of the errors we do when assuming a constant volatility model.

The Gamma has the same value for calls and puts. More precisely,

$$\Gamma Call(S_0, K, 0, T, r, \sigma) = \Gamma Put(S_0, K, 0, T, r, \dot\sigma) = \frac{\mathcal{N}'(d_1)}{S_0 \sigma \sqrt{T}} \quad (6.23)$$

A Python script for Γ is given as follows:

```
def Gamma(S0, K, r, sigma, T):
    #Function to calculate option's gamma
```

The Greeks in the Black-Scholes-Merton Model ■ 135

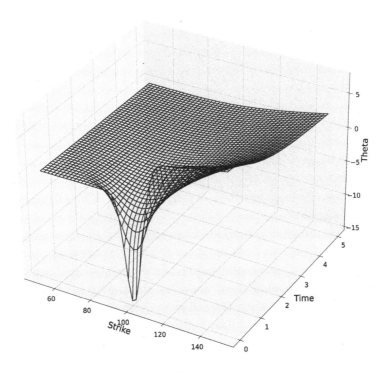

Figure 6.14 Put option theta surface.

```
#S0: Spot price
#K: Strike price
#r: risk-free rate
#sigma: volatility
#T: Time to maturity

#We calculate the probality density func.
pdf=ss.norm.pdf(d1(S0,K,r,sigma,T))

return pdf/(S0*sigma*np.sqrt(T))
```

Notice that Γ is always positive. It takes its maximum for the strike such that $d_1 = 0$, as we can see in Figure 6.15 with parameters $S0 = 100$, $r = 0.05$, $\sigma = 0.2$, $T = 1$

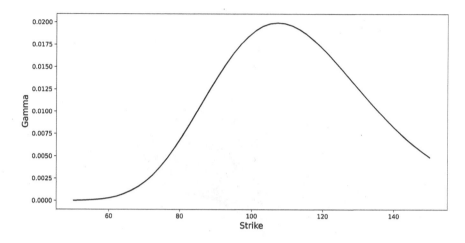

Figure 6.15 Gamma as a function of the Strike.

Moreover, as d_1 is strongly dependent on $\ln S_0 - \ln K$ for short maturities, while this dependence is not so important for long maturities. We can see it empirically with the following parameters $S0 = 110/100/90$, $K = 100$ $r = 0.05$, $\sigma = 0.2$ (see Figure 6.16):

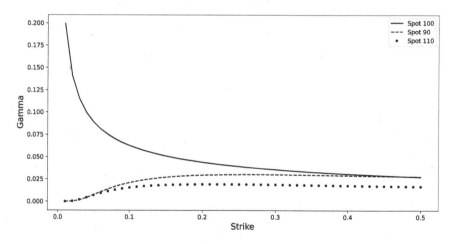

Figure 6.16 Gamma as a function of time to maturity.

We can plot the gamma surface depending on the changes in the strike and the time to maturity. We can observe that options have a higher gamma close to the expiration, especially ATM, but it becomes flat elsewhere (see Figure 6.17).

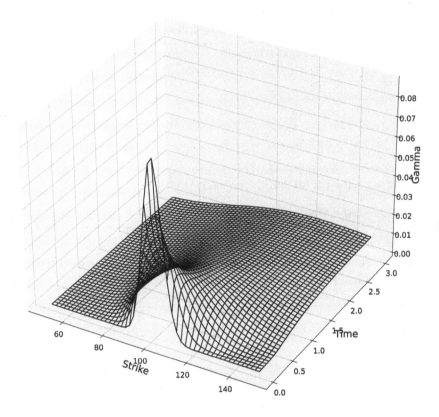

Figure 6.17 Option's Gamma.

6.7.4 Vega

Vega, ϑ, is the derivative of the option price with respect to the volatility parameter.

$$\vartheta Call(S_0, K, 0, T, r, \sigma) = \vartheta Put(S_0, K, 0, T, r, \sigma) = S_0 \mathcal{N}'(d_1)\sqrt{T}. \quad (6.24)$$

Observe that this derivative is **always** positive, and this means that option prices are increasing functions of the volatility σ. Moreover, it is closely related to the Gamma, in the sense that

$$\vartheta Call(S_0, K, 0, T, r, \sigma) = S_0^2 T \sigma \Gamma Call(S_0, K, 0, T, r, \sigma). \quad (6.25)$$

138 ■ A Continuous-time Pricing Model

Vega is very important in volatile markets, it is also one of the hardest as a consequence of the volatility structure on the market.

A Python script for ϑ is given as follows:

```
def vega(S0, K, r, sigma, T):
    #Function to calculate option's vega
    #S0: Spot price
    #K: Strike price
    #r: risk-free rate
    #sigma: volatility
    #T: Time to maturity

    #We calculate the probality density func.
    pdf=ss.norm.pdf(d1(S0,K,r,sigma,T))

    return pdf*S0*np.sqrt(T)
```

We can see that on the short-term, ATM options have higher vega. But as time to maturity increases, vega of OTM options increases faster, having a bigger vega on long maturities (see Figure 6.18).

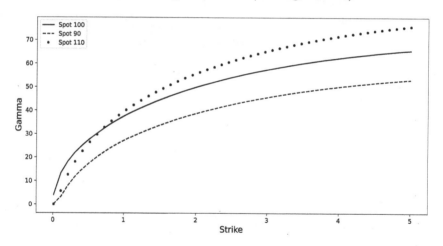

Figure 6.18 Vega as a function of time to maturity.

We can trace the vega surface as a function of strike changes and maturity time. We can see that in the short term it has a symmetric shape that has a peak when the options are ATM. Over time, the symmetry disappears, having instead a curvature (see Figure 6.19).

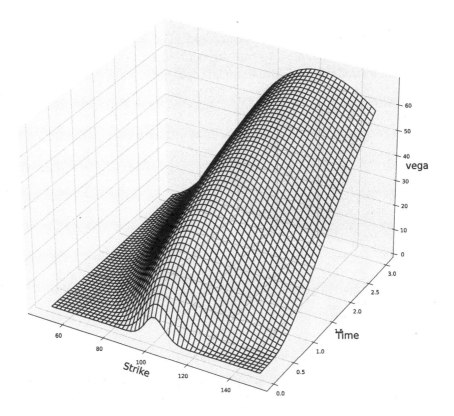

Figure 6.19 Option's vega.

6.8 OTHER ASSETS

Remember that underneath the models, we know that the average scenario must be the Forward value in a risk-neutral measure. That means the classic Black-Scholes-Merton model works for an asset that does not pay dividends. All the formulas following have the same framework but are slightly different.

6.8.1 Black-Scholes-Merton with Dividends

As we suppose in the Forward/Futures calculation, we assume that an asset pays a continuous dividend yield of q. In this case, we have the following Black-Scholes-Merton variation formula.

The formula for a European Call option is

$$Call(S_0, K, t, T, r, \sigma) = e^{-q(T-t)} S_0 \mathcal{N}(d_1) - K e^{-r(T-t)} \mathcal{N}(d_2),$$

where $\mathcal{N}$ denotes the normal distribution function,

$$d_1 = \frac{\ln\left(\frac{S_0}{K}\right) + (r-q)(T-t)}{\sigma\sqrt{T-t}} + \frac{\sigma\sqrt{T-t}}{2}$$

and

$$d_2 = d_1 - \sigma\sqrt{T-t}.$$

In a similar way, the formula for a European Put option is

$$Put(S_0, K, t, T, r, \sigma) = K e^{-rT} \mathcal{N}(-d_2) - e^{-q(T-t)} S_0 \mathcal{N}(-d_1). \quad (6.26)$$

6.8.2 Black-Scholes-Merton for Foreign-Exchange

In the Foreign Exchange case, the extension of the Black-Scholes-Merton model was introduced by Garman and Kohlhagen in 1983 in (Garman and Kohlhagen, 1983).

The formula for a European Call option is

$$Call(S_0, K, t, T, r, \sigma) = e^{-r_f(T-t)} S_0 \mathcal{N}(d_1) - K e^{-r_d(T-t)} \mathcal{N}(d_2),$$

where $\mathcal{N}$ denotes the normal distribution function, r_f is the foreign risk-free rate and r_d is the domestic risk-free rate,

$$d_1 = \frac{\ln\left(\frac{S_0}{K}\right) + (r_d - r_f)(T-t)}{\sigma\sqrt{T-t}} + \frac{\sigma\sqrt{T-t}}{2}$$

and

$$d_2 = d_1 - \sigma\sqrt{T-t}.$$

In a similar way, the formula for a European Put option is

$$Put(S_0, K, t, T, r, \sigma) = K e^{-r_d T} \mathcal{N}(-d_2) - e^{-r_f(T-t)} S_0 \mathcal{N}(-d_1). \quad (6.27)$$

6.8.3 Black-scholes-Merton for Futures

In 1976, Fisher Black made an extension of the Black-Scholes-Merton formula in (Black, 1976) for when the underlying asset is a Future or a Forward.

The formula for a European Call option is

$$Call(S_0, K, t, T, r, \sigma) = e^{-r(T-t)}\left[F\mathcal{N}(d_1) - K\mathcal{N}(d_2)\right],$$

where $\mathcal{N}$ denotes the normal distribution function,

$$d_1 = \frac{\ln\left(\frac{F}{K}\right)}{\sigma\sqrt{T-t}} + \frac{\sigma\sqrt{T-t}}{2}$$

and

$$d_2 = d_1 - \sigma\sqrt{T-t}.$$

In a similar way, the formula for a European Put option is

$$Put(S_0, K, t, T, r, \sigma) = e^{-r(T-t)}\left[K\mathcal{N}(-d_2) - F\mathcal{N}(-d_1)\right]. \quad (6.28)$$

6.9 DRAWBACKS OF THE BLACK-SCHOLES-MERTON MODEL

The Black-Scholes-Merton model is the most famous option pricing model. On the one hand, it is the pillar of modern option pricing. On the other hand, practitioners still use it to quote option prices. But, it is just a model. One useful. But all models have limitations and drawbacks. The Black-Scholes-Merton model too. We review the main ones.

- **Trading cost.** There are no trading costs or fees.

- **Normality.** The Black-Scholes-Merton model assumes that log-prices follow a normal distribution. It is a good idea from which we inherit excellent properties, but it is not completely realistic. See Figure 6.2. One of the most cited problems is that the extremes of the distribution, which we call the tails, are too thin. In other words, the probability of large movements is small. On the other hand, the series of returns exhibit heavy tails, implying that the probability of large movements is not so negligible. See Figure 6.20.

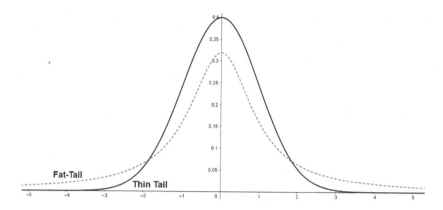

Figure 6.20 Comparison Fat-Tail vs Thin Tail.

- **The Volatility is constant.** The volatility is a constant parameter. The log-returns shows that it is similar to a noise signal. But, if look carefully we will see that at some moments the intensity of the noise change and it changes by periods. This is what we call volatility clustering (see Figure 6.21).

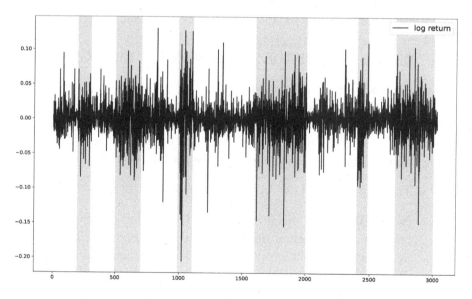

Figure 6.21 Volatility clustering.

6.10 CHAPTER'S DIGEST

In this chapter, we have introduced the Black-Scholes-Merton model. A continuous-time model. It is the most popular model, and although the model was published in 1973, it is still used in practice. We have seen three different views of the model:

- An intuitive derivation from a statistical point of view.

- A replication derivation based on hedging of an option obtaining a Partial Differential Equation.

- The integration of the conditional expectation of the payoff using some probability tools.

In addition, we have introduced two new mathematical structures

- Brownian Motion. A random process with continuous noisy paths and independent increments following a normal distribution.

- Stochastic Differential Equations. A type of differential equation with a deterministic part and a stochastic one.

Beyond that, we have observed that when we have infinity time steps in a Binomial model, it converges to the Black-Scholes-Merton model.

Similar to Chapter 5, we wonder about the Black-Scholes-Merton formula sensitivities. In this case, the Greeks have analytical formulae which allow us to understand the behavior in general cases.

Furthermore, the model has to be coherent with the risk-neutral framework matching the forward value on the average scenario. For that reason, slightly different alternatives to the Black-Scholes-Merton formula have been introduced depending on the underlying type.

Last, but not least, we have presented the main drawbacks and limitations of the Black-Scholes-Merton formula.

6.11 EXERCISES

1. Choose the most correct one. Under Black-Scholes-Merton model:

    (a) $\mathbb{E}(S_T) = S_0 \exp(rT)$.
    (b) $\mathbb{E}(S_T) = S_T$.
    (c) $\mathbb{E}(S_T) = S_0$.

2. Choose the most correct one. Given a close to expiry European call ATM option:

    (a) The Delta is close 1.
    (b) The Delta is close 0.5.
    (c) The Delta is close -1.

3. Choose the most correct one. Given a European Call and Put option with the same maturity and strike:

    (a) The two options have the same vega.
    (b) The two options have the opposite vega.
    (c) It is not possible to know.

4. Assume a non-paying dividend stock such that $S_0 = 100$, r $= 1\%$ (continuously compounded), $T = 1$ and $K = 100$. If the corresponding call option price is 12.368, what is the corresponding implied volatility?

    - 0.3 (30%)
    - 0.2 (20%)
    - 0.1 (10%)

5. Assume that $S_0 = \$100$, $r = 1\%$ (continuously compounded) and that $\sigma = 20\%$. What is the corresponding Black-Scholes-Merton price for an European call with strike price $K = 100$ and time to maturity $T = 1$? What is the price of the corresponding put option? **Solution: c $= 8.4$, p $= 7.4$**

6. What is the price, under the Black-Scholes-Merton model, of a European call option with $S_0 = 100, K = 110, r = 5\%$ (continuously compounded), $\sigma = 30\%$ and $T = 2$? What is the price of the corresponding put option? **Solution: c $= 16.99$, p $= 16.52$**

7. Consider the same data as in Exercise 1. Compute the corresponding call option price assuming a 1-step binomial model. Compare the result with the corresponding Black-Scholes-Merton price. What do you observe? Do the exercise again assuming a 2-steps binomial model and a 3-steps binomial model. **Solution: With one step, c $= 10.41$. With two steps c $= 7.53$, with 3 steps c $\doteq 7.58$.**

8. What it is the probability of a Put Option to end ITM?

9. Prove the Black-Scholes-Merton formula for a Put Option.

CHAPTER 7

Monte Carlo Methods

In the previous chapters, we have seen two different methodologies to obtain the price of an option: the Binomial model and the Black-Scholes-Merton framework. Unfortunately, sometimes the derivative complexity makes it difficult or impossible to use any of these methods. Here, we present an alternative, the Monte Carlo method. This method is computationally intensive. But it is also very flexible since we can use it to price any derivative type.

7.1 THE NEED OF GENERAL OPTION PRICING TOOLS

We have seen in the previous chapters that the price of a European option can be computed as the discounted risk-neutral expectation of its payoff. That is,

$$V = e^{-rT}\mathbb{E}\left[A_T\right], \qquad (7.1)$$

where V is the option price, r denotes the interest rate, T is time to maturity, $\mathbb{E}$ denotes the risk-neutral expectation, and A_T is some random variable that depends on the type of option. For example, A_T can be

- a function of the value at maturity S_T of the underlying asset (as in the case of vanilla options), or

- a random variable that depends on the path of the asset price $\{S_t, t \in [0,T]\}$ from now to maturity (as in the case of path-dependent options as a barrier, a lookback, or an Asian options), or

- a random variable that depends on the price at maturity of a set of assets $\{S_T^1, S_T^2, ...\}$ (as in the case of spread options and basket options).

Then one might think that the valuation of option prices is easily derived from (7.1). For example, in the European call option case with strike K under the Black-Scholes-Merton model, $A_T = \max(S_T - K, 0)$. Then, as we know that S_T follows a lognormal distribution, (7.1) simply

$$V = e^{-rT} \int_{\mathbb{R}} \max(x - K, 0) f(x) dx, \qquad (7.2)$$

where f denotes the lognormal density of S_T, and the problem reduces to the computation of an integral (see again Chapter 6).

In general, it is not straightforward to write the random variable A_T as a function of a random variable with a known density function, even when we assume the Black-Scholes-Merton model. Consider, for example, the case of an Asian option with a payoff

$$\max\left(\sum_{i=1}^{n} S_{\frac{iT}{n}} - K, 0\right),$$

where K is the strike. Here, A_T is a function of the average asset price $\sum_{i=1}^{n} S_{\frac{iT}{n}}$, whose density is unknown.

This example shows us that usually it is not possible to calculate the expectation (7.1), but rather an exception. Even assuming that asset prices follow the Black-Scholes-Merton model, in general, it is not easy to obtain option prices.

Obviously, in the financial industry, it is necessary to calculate expectations such as (7.1) for a wide variety of random variables A_T. Some of them will be more complex and others modeled by more advanced models than Black-Scholes-Merton. For that reason, we need very general tools to calculate (7.1) even when we have no prior knowledge of the density of A_T.

There are several different strategies to calculate options prices, the most popular methodology is given by the so-called Monte Carlo methods, based on simulations. We introduce these methods in the next section.

7.2 MATHEMATICAL FOUNDATIONS OF MONTE CARLO METHODS

Monte Carlo methods are based on simulations. Suppose we flip a coin 10 times. How many times would you expect to get heads? Intuitively, we would say that 5. If we assign tails = 0 and heads = 1, we would say (again, intuitively) that the empirical mean of the experiment has to be around 0.5. What happens if we flip the same coin 1000 times? We would continue to expect the empirical mean to be close to 0.5. Intuitively, we would expect the second experiment to be closer to the theoretical mean than the first. We can confirm this intuition with the following example.

Example 7.1

Let us create a Python script to confirm our intuition. The first step is to define a function called *function_samplemean* that depends on the number of simulations n. The code will flip a fair coin, store all the results in a vector F, and calculate the corresponding mean. We can define this function thanks to the help of the Binomial function implemented in Python under the instruction $np.random.binomial\,(1, 0.5, n)$.

```
#We import the numpy package
import numpy as np

#We define the function sample mean
def function_samplemean(n):
    #We store the results of n Binomial samples
    F=np.random.binomial(1,0.5,n)

    #We calculate and return the mean of the Binomial samples
    return np.mean(F)
```

Now, we launch 10 coins 5 times, or what is the same, we execute the code 5 times choosing $n = 10$.

```
for i in range(0,5):
    print (function_samplemean(10))
```

0.2
0.6
0.4
0.6
0.5

Note that each time the code runs, different values will appear.

We repeat the same experiment, but this time we are going to toss 1000 coins. To do that, we set $n = 1000$.

```
for i in range(0,5):
    print (function_samplemean(1000))
```

```
0.525
0.483
0.525
0.512
0.514
```

As we think, the sample mean is once again around 0.5, but the variability of the results is clearly less.

The above example is a particular case of a more general result. Under some conditions, when we simulate n times a random variable A, the corresponding sample mean tends to be closer to the theoretical mean $\mathbb{E}[A]$. Then, we can compute expectations like in (7.1) by just simulating the payoff and computing the corresponding empirical mean. Before applying this idea to different types of options, let us review the mathematical foundations of this method.

7.2.1 Sample Means as Estimators of Theoretical Expectations

Assume, we simulate n times a random variable X with finite expectation μ and variance σ. This simulation is, mathematically, a sequence of independent and identically distributed random variables $\{X_1, X_2, ...\}$ with the same mean and variance that X. Define now the sample mean $\bar{X} = \frac{1}{n}\sum_{i=1}^{n} X_i$. Then, a direct computation gives us that

$$\mathbb{E}\left[\bar{X}\right] = \frac{1}{n}\sum_{i=1}^{n}\mathbb{E}[X_i] = \frac{1}{n}\sum_{i=1}^{n}\mu = \mu. \tag{7.3}$$

That is, the expectation of the sample mean is μ. Then we say that $\bar{X}$ is an *unbiased* estimator of μ.

On the other hand, as the random variables $\{A_1, A_2, ...\}$ are independent, we get

$$Var(\bar{X}) = \frac{1}{n^2}\sum_{i=1}^{n} Var(X_i) = \frac{1}{n^2}\sum_{i=1}^{n}\sigma^2 = \frac{\sigma^2}{n}. \tag{7.4}$$

Notice that this means that, as n increases, the variance of $\bar{X}$ decreases, as we observed in Example 7.2. Then we say that the sample mean is a more *efficient* estimator of μ when the sample mean increases.

If n is 'big enough', we can think that $\bar{X}$ has to be close to μ, since $\bar{X}$ has an expectation equal to μ and a smaller variance, and this is the basis of Monte Carlo methods.

In fact, we have more information about the behavior of $\bar{X}$. Some basic mathematical results, as the laws of Large Numbers and the Central Limit Theorem, can give more precise information about the behavior of sample means.

7.2.2 The Laws of Large Numbers

The laws of Large Numbers are mathematical results that state that (under some conditions) when we simulate random variables, their sample mean tends (in some mathematical sense) to the corresponding theoretical expectation of this sample mean. In the literature, these theorems are classified into two groups:

- Strong law of Large Numbers
- Weak law of Large Numbers

What is the difference? In the strong law of Large Numbers the convergence is *almost sure*. That means that, with probability one, if we start doing simulations, the sample mean will tend to the theoretical expectation. In the weak law of Large Numbers the convergence is *in probability*. That means that the probability of being near the theoretical expectation tends to one as the number of simulations increases.

These two concepts are illustrated in Figures 7.1 and 7.2..

Example 7.2

In this example, we toss a coin 100 times (i.e., 100 simulations) and see the sample mean evolution's. We can observe that the sample mean tends to be 0.5 (almost sure convergence).

```
#We import the numpy package
import numpy as np

#We import the pyplot package
from matplotlib import pyplot as plt

#We create a vector of 1.001 positions full of zeros
samplemean=np.zeros([1001])

#When we do 0 tosses, the sample mean is 0.
```

152 ■ Monte Carlo Methods

```
samplemean[0]=0

#We do 1.000 tosses and store them in the coin variable
coin=np.random.binomial(1,0.5,1000)

#We calculate the evolution of the sample mean
for i in range (0,1000):
    samplemean[i+1]=samplemean[i]*i/(i+1)+coin[i]/(i+1)

#We plot the evolution of the sample mean
plt.plot(samplemean[1:])
```

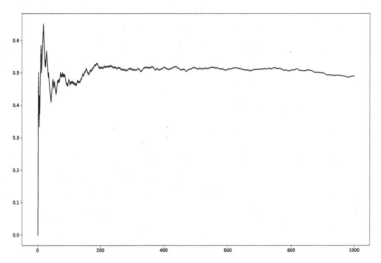

Figure 7.1   Almost sure converge.

Example 7.3

In this example, we repeat the experiment. In this case, we see that the number of paths that are near 0.5 increases with the number of simulations. That is the concept of convergence in probability.

```
#We import the numpy package
import numpy as np

#We import the pyplot package
from matplotlib import pyplot as plt

#We will repeat 200 times the experiment
for i in range (0,200):
```

```
#We create a vector with 101 positions
samplemean=np.zeros([101])

#When we do 0 tosses, the sample mean is 0.
samplemean[0]=0

#We do 100 coin tosses
coin=np.random.binomial(1,0.5,100)

#for each toss, we calculate the evolution of the sample mean
for i in range (0,100):
    samplemean[i+1]=samplemean[i]*i/(i+1)+coin[i]/(i+1)

#We plot each evolution
plt.plot(samplemean[1:])
```

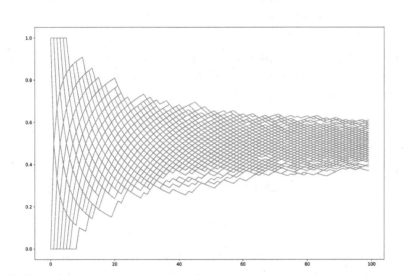

Figure 7.2 Convergence in probability.

These two types of convergence are related to each other. In particular, almost sure convergence implies convergence in probability. All the examples in this book consider random variables with the same (finite) mean. Under this condition, the following version of the Law of Large Numbers (see (Ross, 2014)) guarantees almost sure convergence and, consequently, convergence in probability.

Theorem 7.2.1 (Strong Law of Large Numbers) *Consider a sequence of random variables $\{X_1, X_2, ...\}$ independent and identically* distributed, with mean $\mathbb{E}\left[X_{i}\right]=\mu<\infty$, for all $i \geq 1$. Then

$$\frac{1}{n}\sum_{i=1}^{n}X_{i} \to \mu,$$

almost surely.

We remark that this result guarantees that increasing n 'will get' closer to μ, but is says nothing about the sample size n we need to take to be 'close enough' to μ.

7.2.3 The Central Limit Theorem

The Central Limit Theorem is a convergence *in distribution* result. In particular, it states that, under some conditions, the distribution of the sample mean $\bar{X}$ tends to the normal distribution as n increases. In its classical form, this theorem states as follow.

Theorem 7.2.2 (Central Limit Theorem) *Consider a sequence of independent and identically distributed random variables* $\{X_1, X_2, ...\}$, *with mean* $\mathbb{E}\left[X_{i}\right]=\mu<\infty$ *and* $Var(X_i) = \sigma^2 < \infty$, *for all* $i \geq 1$. *Then*

$$\sqrt{n}(\bar{X} - \mu) \to \mathcal{N}(0,1),$$

as $n \to \infty$.

In particular, this means that, for n big enough, $\bar{X}$ follows approximately a Gaussian distribution with mean μ and variance $\frac{\sigma^2}{n}$. We can observe this phenomenon in the following example.

Example 7.4

In this example, we take $n = 10.000$ and, with this sample size, we execute 100.000 times the function *function_samplesize* as in Example 7.2. Then, we plot the histogram corresponding to this 100.000 realizations of the sample mean. The corresponding python script is given by

```
#We import the numpy package
import numpy as np

#We import the pyplot package
from matplotlib import pyplot as plt

#We set the variable n with the value 10.000
n=10000
```

```
#We create a vector of 100.000 positions full of zeros
simulation=np.zeros(100000)

#We calculate the empirical mean of 10.000 Binomials
#and repeat it 100.000 times.
for i in range (0,100000):
    simulation[i]=function_samplemean(n)

#We plot the histogram with 50 buckets
plt.hist(simulation, 50)
```

The obtained histogram is in Figure 7.3:

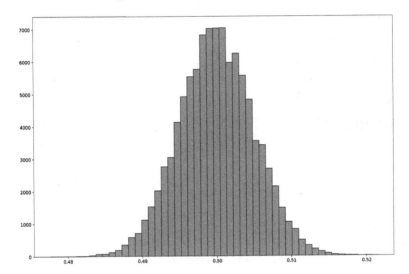

Figure 7.3 Distribution of the empirical mean of 10.000 Binomial.

We observe that this histogram fits a Gaussian bell, according to the Central Limit Theorem.

Once we have seen the mathematical basis of Monte Carlo methods, we study how to apply these simulation methods to different scenarios.

7.3 OPTION PRICING WITH MONTE CARLO METHODS

Option pricing via Monte Carlo methods is based on the simulation of the payoff. This payoff can depend on the final value of the asset, or the path of the asset price, or the final price of two assets, etc. Once we simulate the payoff, we compute the present value of the corresponding sample mean, and this would be our estimation of the option price. Let us see through examples how to apply Monte Carlo techniques in the context of the Black-Scholes-Merton model. The same techniques can be extended to more complex models and are widely used in the financial industry.

7.3.1 European Options that Depend Only on the Final Value of the Asset

Consider the case of a European call option with maturity T and strike price K. In this case, the option price is given by

$$e^{-rT}\mathbb{E}\left[\max(S_T - K, 0)\right], \qquad (7.5)$$

where r is the continuously compounded interest rate and $\mathbb{E}$ denotes the *risk-neutral* probability. In this case, if S follows a Black-Scholes-Merton model, the option price can be explicitly computed as explained in Chapter 6 and no simulation tools are required. Nevertheless, we use this scenario as a toy example to explain the basis of Monte Carlo methods.

In order to estimate (7.5) via Monte Carlo simulation, the detailed steps are the following.

1. Simulate the asset price several times.
2. Compute the corresponding payoffs.
3. Take the mean of the payoffs.
4. Compute the present value.

Let us see a simple Python script to price a European call option with a Monte Carlo. Assume we have a sample size n, first we simulate a vector of n standard random variables z with the instruction *np.random.normal(0,1,n)*. As S follows a Black-Scholes-Merton model, we can simulate n times the random variable S_T by writing

$$S_T = S_0 \exp\left(\left(r - \frac{\sigma^2}{2}\right)T + \sigma\sqrt{T}z\right) \qquad (7.6)$$

(see Chapter 6). Then, we compute the corresponding n payoffs, afterwards, we calculate the mean of the payoffs and discount it to obtain the present value.

Option Pricing with Monte Carlo Methods ■ 157

```python
#We import the numpy package
import numpy as np

#We define the Monte Carlo function
def MonteCarlo(S0,K,r,sigma,T,n):

    #Inputs
    #S0: Asset price
    #K: Strike
    #r: free-interest rates
    #sigma: volatility
    #T: Time to maturity
    #n: Number of simulations

    #We create the vector payoff with n positions full of zeros
    payoff=np.zeros(n)

    #We create the vector S with n positions full of zeros
    S=np.zeros(n)

    #We generate n different random numbers following a N(0,1).
    z=np.random.normal(0,1,n)

    #We calculate the stock value at maturity (it is a random variable).
    S=S0*np.exp(T*(r-sigma**2/2)+sigma*np.sqrt(T)*z)

    #We calculate the payoff
    payoff=np.maximum(S-K,0)

    #We return the discunted payoff
    return np.exp(-r*T)*np.mean(payoff)
```

We illustrate this procedure in the following numerical example.

Example 7.5

Consider the case of a European call option with $S_0 = 100, K = 100, T = 1, r = 0.05$, and $\sigma = 0.2$. A direct computation using the Black-Scholes-Merton formula gives us that the price of this option is equal to 10.45.

On the other hand, we can estimate this option price by using the function Monte Carlo defined in the previous script, for different values of n. Taking $n = 10^i$, for $i = 1, .., 6$, we obtain the following numerical results.

```python
for i in range(1,7):
```

```
print("N. Simulations=", 100**i, "MC estimated option price=",\
MonteCarlo(100,100,0.05,0.2,1,10**i))
```

```
N. Simulations= 100 MC estimated option price= 8.706945098222196
N. Simulations= 10000 MC estimated option price= 9.988705922000955
N. Simulations= 1000000 MC estimated option price=
    10.723374293786922
N. Simulations= 100000000 MC estimated option price=
    10.641472950429424
N. Simulations= 10000000000 MC estimated option price=
    10.517950844804925
N. Simulations= 1000000000000 MC estimated option price=
    10.453662108939815
```

We observe that, as n increases, the Monte Carlo price tends to the true price 10.45.

We notice that, obviously, for other European payoffs depending only on the final value of the asset, the procedure is the same: we would only need to change the call payoff by the corresponding one.

What about American options? In the Binomial model, we have learned to value American options naturally. In the case of the Monte Carlo method, it is also possible, but not straightforward. For this, the Longstaff-Schwartz method, presented in (Longstaff and Schwartz, 2001), must be used. This method is also known as the Least Squares Monte Carlo method. The idea is to simulate the evolution of the assets with a Monte Carlo method, then, from maturity to the beginning, iterating backward. The Least Squares is used at each time step to estimate the expected conditional payoff of continuation. The details of this method are beyond the scope of the book.

7.3.2 European Options that Depend on the Path of Asset Prices

For the sake of simplicity, consider here the case of the same call knock-out option with barrier $b > S_0$. Similar ideas apply for other barrier options, lookback options or Asian options.

In this case, the payoff is given by

$$\mathbb{E}\left[\max(S_T - K, 0)\right], \text{if } S_{max} \leq b, \tag{7.7}$$

and 0 otherwise, where T denotes time to maturity and K is the strike price. Now, let us estimate the option price by a Monte Carlo method.

Assuming a Black-Scholes-Merton model for the asset price S, the steps will be

- Simulate the path of a Brownian motion several times.
- Compute the corresponding paths for the asset price.
- Compute the corresponding payoffs.
- Calculate the mean of the payoffs.
- Compute the present value.

In order to simulate n times a Brownian motion, we have to simulate (n times) the values of this Brownian motion at different time steps, from 0 to T. The number of time steps has to be big enough. For example, we can take this number equal to 100. Then, we have to simulate the matrix numbers

$$W_0^1, W_{\frac{T}{100}}^1, W_{\frac{2T}{100}}^1, ..., W_{\frac{99T}{100}}^1, W_T^1,$$

$$W_0^2, W_{\frac{T}{100}}^2, W_{\frac{2T}{100}}^2, ..., W_{\frac{99T}{100}}^2, W_T^2,$$

$$...$$

$$W_0^n, W_{\frac{T}{100}}^n, W_{\frac{2T}{100}}^n, ..., W_{\frac{99T}{100}}^n, W_T^n,$$

where W^i denotes the i-th simulation of a Brownian motion, $i = 1, ..., n$.

How can simulate all these numbers?

Let us focus on one row. Because of the definition of Brownian motion, we have that

$$W_0^i = 0,$$

and, for all $j = 1, ...100$

$$W_{\frac{(j+1)T}{100}}^i = W_{\frac{jT}{100}}^i + \sqrt{\frac{T}{100}} z,$$

where z denotes a standard normal variable. Then, the first step in our simulation will be the simulation of these $n \times 1.000$ random variables z. Then we compute the n simulations of the Brownian motion path, and then we use the expression

$$S_t = S_0 \exp\left(\left(r - \frac{\sigma^2}{2}\right)t + \sigma W_t\right)$$

(see Chapter 6) to get n simulations of the asset price path.

It is also possible to re-write (7.6) to express it in a recursive way for two consecutive times t_i and t_{i-1},

$$S(t_i) = S(t_{i-1}) \exp\left(\left(r - \frac{\sigma^2}{2}\right)(t_i - t_{i-1}) + \sigma W_{t_i - t_{i-1}}\right). \qquad (7.8)$$

Using Brownian motion properties, it is equivalent to write

$$S(t_i) = S(t_{i-1}) \exp\left(\left(r - \frac{\sigma^2}{2}\right)(t_i - t_{i-1}) + \sigma \sqrt{t_i - t_{i-1}} z_i\right) \qquad (7.9)$$

where z_i, $i = 1, ..., n$ are independent standard normal distributions.

A simple Python script using a Monte Carlo to price a Barrier option.

```
#We import the numpy package
import numpy as np

def MonteCarlo_Barrier(S0,K,r,sigma,T,b,n):

    #Inputs
    #S0: Asset price
    #K: Strike
    #r: free-interest rates
    #sigma: volatility
    #T: Time to maturity
    #b: Barrier Up and out
    #n: Number of simulations

    #We generate a matrix of normal nx101 random numbers
    z=np.random.normal(0,1,[n,100])

    #We create a matrix of nx101 to store the Brownian motion
    brownian=np.zeros([n,101])

    #We create a matrix of nx101 to store the Asset prices
    S=np.zeros([n,101])

    #We create a vector of n positions to store the Payoff
    payoff=np.zeros(n)

    #We set the initial value of the stock to the first column of
    #the Asset Prices matrix
    S[0,:]=S0

    #For each simulation
```

```
for i in range(0,n):

    #For each timestep (we split T in 101 times)
    for j in range(1,101):

        #We generate a Brownian motion of the timestep j*T/101
        brownian[i,j]=brownian[i,j-1] + np.sqrt(T/101)*z[i,j-1]

        #We generate the asset path
        S[i,j]=S0*np.exp((r-0.5*sigma**2)*(j*T/101)\
                +sigma*brownian[i,j])

    #Once we have the full path, look if the barrier is reached
    #if so, the option is deactivate it
    if np.max(S[i,:])>b:
        payoff[i]=0

    #If not, we have a European option
    else:
        payoff[i]=np.maximum(S[i,100]-K, 0)

#We discount the Payoff to move from T to now.
return np.exp(-r*T)*np.mean(payoff)
```

In this example, we have divided the time to option expiration into 100 intervals to see if the barrier is reached. That is an arbitrary number. It is possible to divide it into as many intervals as you like. Increasing the intervals will increase the calculation time.

We can illustrate this procedure with the following example.

Example 7.6

Consider knock-out barrier options with payoff (7.7) and take $S_0 = 100, K = 100, T = 1, r = 0.05, \sigma = 0.2$, and for different barriers $b = 110, 120, ..., 200$. A Monte Carlo computation of the corresponding price, with a number of simulations $n = 100.000$ gives the following results:

```
for i in range(0,10):
    print ("Barrier=", 110+10*i, "Option price=",\
    MonteCarlo_Barrier(100,100,0.05,0.2,1,110+10*i,100000))
```

Barrier= 110 Option price= 0.17613687992433594
Barrier= 120 Option price= 1.422708350409316
Barrier= 130 Option price= 3.6875406499962597
Barrier= 140 Option price= 6.067045018018231

```
Barrier= 150 Option price= 7.854035576434033
Barrier= 160 Option price= 9.019777871071692
Barrier= 170 Option price= 9.670969671827839
Barrier= 180 Option price= 10.041875886799316
Barrier= 190 Option price= 10.262901117595776
Barrier= 200 Option price= 10.25878510253609
```

As expected, the barrier option price is less than the corresponding vanilla price (with no barrier), which was equal to 10.45 (see Example 7.5). Moreover, when the barrier increases, the barrier option price tends to the vanilla price.

The key point in the above script was the simulation of the Brownian motion and the asset price paths. The same procedure in this example can be applied to other kinds of path-dependent options, like lookback or Asian options, by simply changing the payoff.

7.4 EUROPEAN OPTIONS THAT DEPEND ON THE FINAL PRICE OF TWO ASSETS

Consider, for example, the case of a spread option with payoff

$$e^{-rT}\mathbb{E}\left[\max(S_T^2 - S_T^1 - K, 0)\right], \quad (7.10)$$

for some maturity T and some strike K, and where we consider that both S^1, S^2 follow (under a risk-neutral probability) a Black-Scholes-Merton model. That is, they admit a representation

$$S_T^1 = S_0^1 \exp\left(\left(r - \frac{(\sigma^1)^2}{2}\right) + \sigma^1\sqrt{T}z_T^1\right), \quad (7.11)$$

$$S_T^2 = S_0^1 \exp\left(\left(r - \frac{(\sigma^2)^2}{2}\right) + \sigma^2\sqrt{T}z_T^2\right), \quad (7.12)$$

where z^1, z^2 are two standard Gaussian random variables that may be correlated. As discussed in Chapter 4, spread options are habitual in commodity markets. In this scenario, one asset is typically the 'input' and the other the 'output' of a production process, such as crude oil and gasoline. In this context, we expect asset prices to be correlated, which implies that z^1, z^2 are correlated.

The key point in the simulation of S^1, S^2 is the simulation of the random variables z^1, z^2. Once we simulate them, equations (7.11) and (7.12) allow to directly simulate the values of S^1 and S^2 at maturity.

A simple way to simulate z^1, z^2 consists of the following steps:

- Simulate two independent standard random variables z^1, u.
- Generate the random varible $z^2 = \rho z_T^1 + \sqrt{1-\rho^2} u$,

where $\rho \in [-1, 1]$ is the correlation coefficient. The basic properties of the Gaussian distribution allow us to see that z^2, defined as before, is a standard normal variable and that its correlation with z^1 is equal to ρ.

Once we simulate z^1 and z^2 we can compute the option price of options depending on these two underlyings. In particular, we can compute the spread option price with payoff (7.10).

A simple Python script for this problem is given by

```
#We import the numpy package
import numpy as np

def MonteCarloSpread(S01, S02, K, r, sigma1, sigma2, rho, T, n):

    #Inputs
    #S01: Asset price 1
    #S02: Asset price 2
    #K: Strike
    #r: free-interest rates
    #sigma1: volatility asset 1
    #sigma2: volatility asset 2
    #rho: Correlation
    #T: Time to maturity
    #n: Number of simulations

    #We create a vector of n positions to store the Payoff
    payoff=np.zeros(n)

    #We create a vector of n positions to store the Asset prices #1
    S1=np.zeros(n)

    #We create a vector of n positions to store the Asset prices #2
    S2=np.zeros(n)

    #We generate a normal variable
    z1=np.random.normal(0,1,n)

    #We generate an auxiliar random variable
```

```
u=np.random.normal(0,1,n)

#We generate a correlated random variable
z2=rho*z1+np.sqrt(1-rho**2)*u

#We generate the values at T of the asset #1
S1=S01*np.exp((r-0.5*sigma1**2)*T + sigma1*np.sqrt(T)*z1)

#We generate the values at T of the asset #2
S2=S02*np.exp((r-0.5*sigma2**2)*T + sigma2*np.sqrt(T)*z2)

#We calculate the payoff
payoff=np.maximum(S1-S2-K, 0)

#We return the discounted payoff to move the Payoff from T to now
return np.exp(-r*T)*np.mean(payoff)
```

We can see a numerical experiment in the following example.

Example 7.7

Consider a spread option with payoff (7.10) and $S_0^1 = S_0^2 = 100$, $K = 0$, $\sigma^1 = 0.3$, $\sigma^2 = 0.2$, and varying values of the correlation parameter ρ. We take the number of simulations $n = 100.000$. We can see the effect of ρ has a strong impact on the spread option price.

```
for i in range(0,10):
    rho=round(0.1*i, 1)
    print ("Correlation=", rho, "Spread Option price=",\
    MonteCarloSpread(100, 100, 0, 0.05, 0.3, 0.2, rho, 1, 100000))
```

```
Correlation= 0.0 Spread Option price= 14.29179835298491
Correlation= 0.1 Spread Option price= 13.594930365905528
Correlation= 0.2 Spread Option price= 12.836185198375233
Correlation= 0.3 Spread Option price= 12.1895145694216
Correlation= 0.4 Spread Option price= 11.442581030661641
Correlation= 0.5 Spread Option price= 10.564239944374384
Correlation= 0.6 Spread Option price= 9.52638889317158
Correlation= 0.7 Spread Option price= 8.590527151250962
Correlation= 0.8 Spread Option price= 7.367997834224422
Correlation= 0.9 Spread Option price= 5.945330840906784
```

7.5 CHAPTER'S DIGEST

A Monte Carlo method is a mathematical technique based on the computation of a sample mean $\bar{X}$, where the sample is obtained by simulation. $\bar{X}$ is an unbiased estimator of the theoretical expectation $\mathbb{E}[X]$, which is more efficient as the number of simulations increase. Then, a simulation of the payoff function becomes a generic tool to compute European option prices.

Classical convergence results include the weak and the strong laws of the Large Numbers. In the weak law the convergence is in probability, while in the strong law the convergence is almost sure. Another classical result, the Central Limit Theorem, states that the sample mean, for n big enough, follows approximately a Gaussian distribution.

As examples, we have implemented a Monte Carlo method for the Black-Scholes-Merton model to price:

- A European call and put option.

- A path-dependent option. In particular, a Barrier option.

- An option depending on more than one underlying assets.

We have seen that in the case of a European call or put option and assuming that S follows a Black-Scholes-Merton model when the number of simulated trajectories goes to infinity, the Monte Carlo price converges to the same price as the one given by the Black-Scholes-Merton formula. Remember that a similar situation happens with the Binomial model when the time steps go to infinity. Nevertheless, the situation is different. The Binomial model is an approximation to the Black-Scholes-Merton model. As a consequence, Binomial prices approximate Black-Scholes-Merton prices. The Monte Carlo method is not a model but a tool that can be applied to different models.

7.6 EXERCISES

1. Compute the Monte Carlo price of a European put with $S_0 = 90, r = 0.02, T = 0.5, \sigma = 0.3$, and $K = 92$, from a simulation of sample size $n = 10$ and assuming the Black-Scholes-Merton model. Do the same exercise 50 times.

    (a) What is the mean of the 50 obtained results? And the standard deviation?

(b) Do the same exercise as before, but now taking $n = 1000$. What do you observe?

(c) Compare the above results with Equations 7.3 and 7.4.

2. Assume the Black-Scholes-Merton model. According to the Strong Law of Large Numbers, the Monte Carlo price of a European call has to tend to the corresponding Black-Scholes-Merton price as the sample size n tends to infinity. Write a Python script to show this phenomenon, in a similar way as we did in Example for the case of a coin. Take $S_0 = 120, r = 0.05, T = 0.8, \sigma = 0.3$ and $K = 120$.

3. Assume the Black-Scholes-Merton model. Write a Python script to compute the Monte Carlo price of a lookback option with payoff $\max(S_{\min} - K, 0)$, where $S_0 = 110, r = 0.02, T = 0.5, \sigma = 0.3$, and $K = 100$. Take different values and, for every chosen n, run 50 times the code. What is the minimum value of n you required to get correct decimal points?

4. Assume the Black-Scholes-Merton model. Write a Python script to compute the Monte Carlo price of an option with payoff $(S_{\max} - S_{\min})$, where $S_0 = 110, r = 0.05, T = 1, \sigma = 0.3$. Take different sample size values and, for every chosen n, run 50 times the code. What is the minimum value of n you required to get correct decimal points?

5. Write a Python script to compute the Monte Carlo price of an option with payoff $(S^1_{\max} - S^2_{\min})$, where S^1, S^2 denote the prices of two assets under the Black-Scholes-Merton model, with $S^1_0 = 110, S^2_0 = 110, r = 0.05, T = 1, \sigma^1 = 0.3, \sigma^2 = 110$, and $\rho = 0.8$. Take different sample size values and, for every chosen n, run 50 times the code. What is the minimum value of n you required to get correct decimal points?

6. We simulate the evolution of a stock with a Monte Carlo with parameters $S_0 = 100$, $\sigma = 30\%$, $r = 5\%$, $T = 1$, $n = 5$ and we obtain the following simulations

Price
105
112.3
103.1
96.6
87.3

    - What it is the corresponding Monte Carlo estimation of the price of an European Put option with $K = 100$? **Solution: 3.06**

CHAPTER 8

The Volatility

In the Black-Scholes-Merton model, the volatility is assumed to be constant. Nevertheless, this is not consistent with real market data. In this chapter, we study the different definitions of volatility and we discuss their use in practice.

8.1 HISTORICAL VOLATILITIES

In the Black-Scholes-Merton model, the volatility is assumed to be a constant. Recall that, under this model, the asset price S follows the equation
$$dS_t = rS_t dt + \sigma S_t dW_t,$$
where r denotes the interest rate and W is a Brownian motion. As pointed out in Chapter 6, the log-prices $X = \ln S$ satisfy that
$$dX_t = \left(r - \frac{\sigma^2}{2}\right) dt + \sigma dW_t$$
which implies that the standard deviation of the log-returns dX_t is given by $\sigma\sqrt{dt}$. This means that, if market prices fit the Black-Scholes-Merton model, the volatility can be estimated from the standard deviation of the asset price returns. More precisely, one should follow the steps:

- take observations $S_0, S_1, .., S_n$ at intervals of dt years (for example, if data is weekly, $dt = 1/52$)

- compute the log-return $u_1, u_2, ..., u_n$ in each interval:
$$u_i = \ln S_i - \ln S_{i-1} = \ln\left(\frac{S_i}{S_{i-1}}\right)$$

DOI: 10.1201/9781003266730-8

170 ■ The Volatility

- compute the standard deviation s of the sample $u_1, ..., u_n$. Then, the volatility estimate $\hat{\sigma}$ is given by $\hat{\sigma} = \frac{s}{\sqrt{dt}}$.

The volatility estimate $\hat{\sigma}$ is called the **historical volatility**. Let us see this computation in the following example.

Example 8.1

Assume that the year has 252 trading days and consider the data in Table 8.1, where we can see daily asset prices in 20 consecutive days (S_i, $i = 0, ...19$) and the corresponding log-returns. The standard deviation of the 9 log-returns is equal to 0.0206485, which implies that the annualized volatility σ is estimated by

$$\hat{\sigma} = \frac{0.0206485}{\sqrt{\frac{1}{252}}} = 0.0206485\sqrt{252} = 0.327785.$$

Nevertheless, if we estimate the volatility in different time periods, we can observe that the estimation differs, as we see in the following section.

Table 8.1 Simulation of daily asset prices and the corresponding returns (in euros).

Day	Asset price	Log-return
0	50.00	—
1	51.443261763565765	0.02845648158048313
2	50.655919121355794	−0.015423399936464216
3	48.32551154424598	−0.047096476534655735
4	48.4674375967083	0.002932572164302162
5	48.61491489508392	0.0030381919686215475
6	49.00550271043332	0.008002217347921598
7	50.911565166993576	0.03815751922597194
8	50.13224433131135	−0.01542571074742565
9	49.68231850630957	−0.009015295154079126
10	49.12051588910042	−0.01137231893082118
11	50.69754192822086	0.03160064033172414
12	51.7271232889061	0.020104845613396538
13	51.27104760516041	−0.008856053667858098
14	51.79188720938269	0.010107300730875587
15	50.675090751532935	−0.021799036229550817
16	51.988409460650985	0.02558631555857843
17	52.821735763781376	0.015901969446244786
18	53.36958956567285	0.010318331764357059
19	52.42790291900512	−0.017802151905467033

8.2 THE SPOT VOLATILITY

If asset prices follow the Black-Scholes-Merton model, the historical volatilities computed in different periods have to be similar, but they differ a lot. Let us see for example the data in Figure 8.1. In this dataset, the volatility of the EURO STOXX 50 has been estimated every day, from an intra-day 5-minutes subsample.

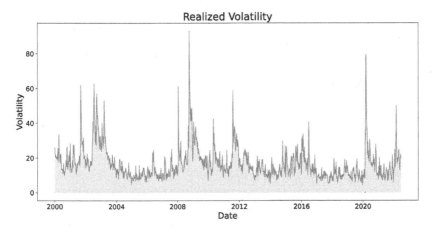

Figure 8.1 Estimated daily volatility (in %) for the EURO STOXX50 from (Heber et al., 2009).

As we can observe, daily volatility is far away from being constant and it behaves as random. This leads us to consider the volatility σ as a random process $\sigma = \{\sigma_t, t \in [0, T]\}$. That is, to extend the Black-Scholes-Merton model allowing the volatility to be random

$$dS_t = rS_t dt + \sigma_t S_t dW_t.$$

Instantaneous volatilities σ_t are called **spot volatilities**. Different choices of the process σ have been proposed in the literature. In **local volatility models**, the volatility is assumed to be a function of time and the underlying asset price. That is,

$$\sigma_t = \sigma(t, S_t),$$

for some function σ. Local volatility models can be calibrated from real market data (see (Dupire, 1994)) and are useful in the computation of European option prices. Nevertheless, these models are not adequate to reproduce the price of path-depended options (see for example (Alòs and García Lorite, 2021)).

Another classical approach is to allow σ to follow another stochastic differential equation, driven by another Brownian motion B, that can be correlated with W. This choice have some advantages but also some drawbacks (see again (Alòs and García Lorite, 2021)). In order to price both European and exotic options, a common approach is to consider **stochastic-local volatility models**, where the volatility is of the form

$$\lambda(t, S_t)\sigma_t,$$

where λ is a deterministic function and σ is a diffusion process.

8.3 THE IMPLIED VOLATILITY

Modeling directly the market volatility is not trivial, since this process is not observed, but estimated. Then, how can we construct an adequate model for option prices? The idea is to find a model reproducing market prices of vanillas, and to use these models to price exotics.

A useful tool for this objective is the **implied volatility**, defined as the volatility that we have to write in the Black-Scholes-Merton formula. That is, the implied volatility I is defined as the quantity satisfying the equation

$$BS(S_0, K, r, x, T) = V(T, K),$$

where $BS(S_0, K, r, x, T)$ is the Black-Scholes-Merton price of a call (or put) with time to maturity T, strike price K, interest rate r, and volatility I, and $V(T, K)$ denotes the corresponding market option price. Notice that, because of the put-call parity relationship, the implied volatility is the same computed from calls or from puts. Moreover the implied volatility is a quantity that depends on T and K, and then we will denote it by $I(T, K)$.

There is not an explicit expression for the implied volatility. Then, $I(T, K)$ has to be approximated by numerical methods. Notice that $I(T, K)$ is, by definition, a minimum of the function f defined as

$$f(x) = (BS(S_0, K, r, x, T) - V(T, K))^2,$$

an expression that allows us to use numerical optimization methods. In particular, we can use Brent's method (see (Brent, 1973)) as we can see in the following example.

Example 8.2

Consider a call option with $S_0 = 100$, $K = 100$, $r = 0.05$, $T = 1$ that has a market price equal to 10. What is the corresponding implied volatility? A direct implementation of the Brent's method in Python gives us the following result:

```
import numpy as np
from scipy import optimize

#We define the error function
def f(x):
    return(BlackScholescall(100,100,0.05,x,1)-25)**2

#We apply Brent's method to find the minimum
#in brack=(0.001,3) we specify an interval
#where we believe the minimum has to be in
print('The estimated IV is', optimize.brent(f,brack=(0.001,3)))
```

The estimated IV is 0.1879716496218057

The plot of the implied volatility as a function of the strike and time to maturity is called the **implied volatility surface** (see Figure 8.2).

Reproducing option prices is equivalent to the replication of the implied volatility surface. Because of technical reasons, this surface (instead of directly option prices) is commonly used in the calibration of pricing models. The calibration of this models is out of the scope of this course. A more detailed introduction to this topic can be found, for example, in (Bergomi, 2016).

Notice that, under the Black-Scholes-Merton model, the volatility is a constant. If this model was true, market implied volatilities would be constant and the implied volatility surface would be flat. But this is not the case. If we fix time to maturity, we can see that the implied volatility exhibits a U-shaped pattern as function of the strikes, a phenomenon that is called the **smile**. When this pattern is asymmetric, we call it a **skew**, or some times, a **skewed smile**. This skews and smiles tend to be more pronounced at short maturities, and they flatten as time to maturity increases, as we can observe in Figure 8.2.

174 ■ The Volatility

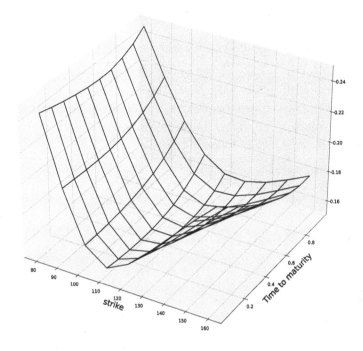

Figure 8.2 Ficticious implied volatility surface as a function of the strike and time to maturity.

8.4 CHAPTER'S DIGEST

The Black-Scholes-Merton model is not enough to reproduce real market data, since it assumes that the volatility is a constant, that can be estimated as the standard deviation of the log-returns.

This estimate of the volatility is called the historical volatility and we can observe that it is not constant, but it changes from time to time. This leads to the modeling of the volatility as a random process. Instantaneous volatilities σ_t are called **spot volatilities**. A useful tool in the calibration of a model for option pricing is the **implied volatility**, that is, the volatility that has to be used in the Black-Scholes-Merton equation to get the option market price. The plot of the implied volatility as a function of the strike and time to maturity is called the **implied**

volatility surface, that exhibits smile and skew effects that flatten with time to maturity.

8.5 EXERCISES

1. Consider the following asset price data. What is the corresponding historical volatility?

Day	Asset price
0	100
1	98.48
2	97.21
3	94.9
4	91.27
5	92.71
6	94.29
7	94.73
8	93.65
9	90.81
10	93.24
11	94.62
12	93.07
13	92.77
14	92.5
15	90.89
16	91.38
17	92.59
18	92.34
19	96.52
20	96.1

   Solution: 0.31

2. What is the implied volatility of an European call with $S_0 = 50$, $K = 55$, $r = 0.02$, $T = 0.5$ and market price equal to 4?
   Solution: 0.406

3. Can you define the implied volatility?

4. Choose the most correct one.

   (a) The implied volatility is computed from asset prices.

(b) In real market data, we observe that implied volatilities depend on the strike and on time to maturity.

(c) The implied volatility is defined as the standard deviation of the returns.

CHAPTER 9

Replicating Portfolios

Our objective in this chapter is to construct a replicating portfolio. That is, given an European option (say, a vanilla call or put option), we want to construct a portfolio with assets and risk-free investment and with the same value (at every moment) that this option.

Is this possible? Only partially. Let us study this replication for the Binomial model and for the Black-Scholes-Merton model.

9.1 REPLICATING PORTFOLIOS FOR THE BINOMIAL MODEL

Let us consider a one-step binomial model. We have seen in Chapter 5 that a portfolio composed of

- An European call or put, and
- $-\Delta$ units of the underlying asset

is risk-free. That is,

$$\text{Option-}\Delta \text{ Assets} = \text{Cash},$$

in the sense that a portfolio composed of an long option and a short position in Δ assets is equivalent (has the same payoff) to a risk-free investment. This observation allowed us to compute the option price in Chapter 5, and the same idea can be applied to the construction of a replicating portfolio. Notice that the above equality can be written as

$$\text{Option} = \text{Cash} + \Delta \text{ Assets},$$

DOI: 10.1201/9781003266730-9

which implies that an option has the same payoff as a portfolio composed of Δ units of the asset and some risk-free investment. Obviously, the amount of this risk-free investment has to be the 'remaining', that is, the value of the option minus the value of Δ units of the asset. Then, the replicating portfolio of an option should be composed of

- Δ units of the underlying asset, and
- A risk-free investment equal to the option price minus the value of the above Δ units of the asset.

If we denote by V_0 the option price and by S_0 the asset price at $t=0$, the price of this portfolio at inception is equal to

$$\Delta S_0 + (V_0 - \Delta S_0) = V_0.$$

At maturity T, the value of the investment in assets is S_T and the value of the risk-free investment is $(V_0 - \Delta S_0)e^{rT}$. Then, the value of this portfolio at maturity is equal to

$$\Delta S_T + (V_0 - \Delta S_0)e^{rT}$$

As the combination of a long option and a short position in Δ assets is risk-free, $V_0 - \Delta S_0 = (V_T - \Delta S_T)e^{-rT}$, which allows us to write the final value of our replicating portfolio as

$$\Delta S_T + (V_T - \Delta S_T) = V_T,$$

which proves that our portfolio replicates the option.

In a binomial model with several steps, the procedure would be similar but, as Δ changes at every node, we should have to 'reorganize' the composition of the replicating portfolio at every step. That is, the global strategy should be

- At $t=0$, consider an amount of money equal to the corresponding option price V_0.

- At $t=0$, construct a portfolio that consists of Δ_0 units of the asset and a risk-free investment equal to $V - \Delta S_0$.

- At every node, we reorganize our portfolio, in such a way that the number of assets equals the corresponding Delta. 'Reorganizing' means that the portfolio is self-financing. That is, we do not add nor we withdraw money from outside.

A replicating strategy where the replicating portfolio is adjusted during the life of the option is called a **dynamic hedging**. When these adjustments are not necessary (as in the case of the one-step binomial model) we refer to this strategy as a **static hedging**. In general, if Δ changes from day to day, the replicating strategy has to be reorganized from day to day, leading to a dynamic hedging.

Let us see a numerical example of a dynamic hedging in a 2-steps binomial model

Example 9.1

Assume a two-step binomial model as in Figure 9.1 and consider a European put option with strike 8€ and interest rate equal to 1% cc.

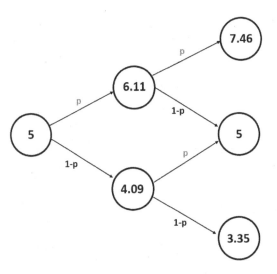

Figure 9.1 Binomial Model: asset prices.

The option price tree is given in Figure 9.1, as well as the corresponding Deltas.

180 ■ Replicating Portfolios

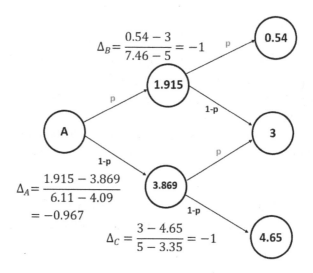

Figure 9.2 Binomial Model in Example 5.7.

At node A, the option price is 4.977 and the Delta is -0.967. Then the replicating portfolio at inception consists of (see Figure 9.2)

- -0.967 assets (that is, a short position in 0.967 assets), and
- a risk-free investment of $4.977 + 0.967 \times 5 = 9.812$.

Then, imagine that the asset price increases and we move to the node B. Then, the value of the option (and then the value of the replicating portfolio) is 1.915, and the Delta value has changed to -1, see Figure 9.2. Then, we have to sell some assets to reorganize the portfolio as

- -1 asset (that is, a short position in one assets), and
- a risk-free investment of $1.915 + 1 \times 6.11 = 8.025$

Notice that, independently on the final value of the asset being 7.46 or 5, the value of the replicating portfolio will be, by construction, the final value (the payoff) of the option.

9.2 REPLICATING PORTFOLIOS FOR THE BLACK-SCHOLES-MERTON MODEL

The Black-Scholes-Merton model can be seen as the limit of a binomial model as the number of steps tends to zero. Then, the replicating strategy would be constructed in a similar way as before, reorganizing the portfolio in such a way the number of assets in this portfolio is always equal to Delta. Nevertheless, as we discussed in Chapter 6, the Delta of an option changes continuously during the life of the option. More precisely

$$\Delta = N(d_1)$$

for a call, and

$$\Delta = N(d_1) - 1$$

for a put, where

$$d_1 = \frac{\ln(S_t/K) - r(T-t)}{\sigma\sqrt{T-t}}.$$

Notice that Delta is positive for calls, and negative for puts. As these Deltas change continuously with time, we should have to **reorganize** our portfolio continuously. In practice, this continuous rebalancing is impossible, and this is done daily, weekly, or when we consider more convenient. Notice that, as the perfect hedging is impossible, there is a **hedging error**, that is, a difference between the value of the replicating portfolio and the value of the option. In particular, we can not assume at every step (except at $t = 0$) that the value of the portfolio is the value of the option.

Let us see the procedure in detail. To fix ideas, let us assume that the option is an European call (we would proceed in a similar way if the option is an European put). Let us denote by N the number of times we decide to act on our replicating portfolio. Then the time interval will be $\delta t = T/N$. We denote RF_i, $Call_i$, Δ_i and $Port_i$ the values of these quantities at time $i = iT/N$, $i = 1, ..., N$. Then

- at $t = 0$:
    - The price of the replicating portfolio, $Port_0$, is the same as the corresponding option price, $Call_0$.
    - Then, the risk-free investment RF_0 is given by

    $$RF_0 = Call_0 - \Delta_0 S_0.$$

- At $t = T/N$ (after one step):
    - The price of this portfolio has changed to
    $$Port_1 = RF_0 \exp(r\delta t) + \Delta_0 S_1.$$
    - Now, we reorganize it. Therefore, the new number of assets will be Δ_1, and then
    $$RF_1 = Port_1 - \Delta_1 S_1.$$
- At $t = 2T/N$:
    - The price of this portfolio has changed again. Now
    $$Port_2 = RF_1 \exp(r\delta t) + \Delta_1 S_2.$$
    - Now, we reorganize it again. So, the new number of assets will be Δ_2, and
    $$RF_2 = Port_2 - \Delta_2 S_2.$$
- and so on...

We can see this procedure in the following example

Example 9.2

Take $r = 0.05$, $\sigma = 0.2$, $S_0 = 100$ and consider a vanilla call with $K = 90$ and $T = 0.5$. Imagine we replicate this call taking $N = 5$.

If $S_{0.1} = 90$, $S_{0.2} = 95$, $S_{0.3} = 100$, $S_{0.4} = 105$ and $S_{0.5} = 100$,

what is the final value of our replicating portfolio? What is the final value of the option?

- At time $t = 0$, $d_1 = 0.992$ and $d_2 = 0.851$. Then both $Call_0$ and $Port_0$ are given by
$$V_0 = 100 N(0.992) - 90 * e^{-0.05 \times 0.5} N(0.851) = 13.498$$

Notice that
$$\Delta = N(0.992) = 0.8394012.$$

Then, the composition of the replicating portfolio at time $t = 0$ is

- Value in assets: $\Delta S_0 = 0.8394012 \times 100 = 83.94012$.
- Value in risk-free investment: $13.498 - 83.94012 = -70.45188$.

- At time $t = 0.1$, after one step, the value of the portfolio has changed. Now its value is

$$\begin{aligned} Port_1 &= \Delta_0 S_1 + RF_0 \exp(0.05 \times 0.1) \\ &= 0.8394012 \times 90 - 70.45188 \exp(0.05 \times 0.1) \\ &= 4.75. \end{aligned}$$

Now, we reorganize it. As at this moment

$$d_1 = \frac{\ln 90 - \ln 90 + 0.05 \times 0.4}{0.2\sqrt{0.4}} + \frac{0.2\sqrt{0.4}}{2} = 0.22$$

and then we get $\Delta_1 = N(0.22) = 0.59$. Then the portfolio consists of 0.59 assets and a risk-free investment of $RF_1 = Port_1 - \$\Delta_1 S_1 = 4.75 - 0.59 \times 90 = -48.13$.

- At time $t = 0.2$, after 2 steps, the value of the portfolio has changed again. Now its value is

$$\begin{aligned} Port_2 &= \Delta_1 S_2 + RF_1 \exp(0.05 \times 0.1) \\ &= 0.59 \times 95 - 48.13 \exp(0.05 \times 0.1) \\ &= 7.44. \end{aligned}$$

Now, we reorganize it. As at this moment

$$d_1 = \frac{\ln 95 - \ln 90 + 0.05 \times 0.3}{0.2\sqrt{0.3}} + \frac{0.2\sqrt{0.3}}{2} = 0.69$$

and then we get $\Delta_2 = N(0.69) = 0.75$. Then the portfolio consists of 0.75 assets and a risk-free investment of $RF_2 = Port_2 - \Delta_2 S_2 = 7.44 - 0.75 \times 95 = -64.13$.

- At time $t = 0.3$, after 3 steps, the value of the portfolio has changed again. Now its value is

$$\begin{aligned} Port_3 &= \Delta_2 S_3 + RF_2 \exp(0.05 \times 0.1) \\ &= 0.75 \times 100 - 64.13 \exp(0.05 \times 0.1) \\ &= 10.89. \end{aligned}$$

Now, we reorganize it. As at this moment

$$d_1 = \frac{\ln 100 - \ln 90 + 0.05 \times 0.2}{0.2\sqrt{0.2}} + \frac{0.2\sqrt{0.2}}{2} = 1.33$$

and then we get $\Delta_3 = N(1.33) = 0.91$. Then the portfolio consists of 0.91 assets and a risk-free investment of $RF_3 = Port_3 - \Delta_3 S_3 = 10.89 - 0.91 \times 100 = -80.00$.

- At time $t = 0.4$, after 4 steps, the value of the portfolio has changed again. Now its value is

$$\begin{aligned} Port_4 &= \Delta_3 S_4 + RF_3 \exp(0.05 \times 0.1) \\ &= 0.91 \times 105 - 80.00 \exp(0.05 \times 0.1) \\ &= 15.04. \end{aligned}$$

Now, we reorganize it. As at this moment

$$d_1 = \frac{\ln 105 - \ln 90 + 0.05 \times 0.1}{0.2\sqrt{0.1}} + \frac{0.2\sqrt{0.1}}{2} = 2.55$$

and then we get $\Delta_4 = N(2.55) = 0.99$. Then the portfolio consists of 0.99 assets and a risk-free investment of $RF_4 = Port_4 - \Delta_4 S_4 = 15.04 - 0.99 \times 105 = -89.39$.

- At time $t = 0.5$, after 5 steps, the value of the portfolio has changed again. Its value at maturity is given by

$$\begin{aligned} Port_5 &= \Delta_4 S_5 + RF_4 \exp(0.05 \times 0.1) \\ &= 0.99 \times 100 - 89.39 \exp(0.05 \times 0.1) \\ &= 9.61. \end{aligned}$$

Notice that, at maturity, the value of the option is equal to 10. Then, the hedging error (that is, the difference at maturity between the option price and the value of the replicating portfolio) is $10 - 9.61 = 0.39$.

A numerical simulation is shown in the following example.

Example 9.3

Let us consider a European call with $S_0 = 100$, $K = 100$, $r = 0.0$, $\sigma = 0.3$, and $T = 1$. A simulation of this option price and the value of the correspondiing replicating portfolio with $N = 10$ from inception to maturity is shown in Figure 9.3. The same evolution is represented in Figure 9.4, but with $N = 100$. Notice that the higher the number of times we reorganize the portfolio, the less the hedging error.

```
#packages needed

#The numerical package
import numpy as np

#The stattiscal package
import scipy.stats as ss

#The graphical package
import matplotlib.pyplot as plt

## Stock Inputs
#Asset Initial Value
S0=100
#Strike value
K=100
#Risk-free interest rate
r=0.0
#Volatility
sigma=0.3
#Time
T=1
#Number of time steps
N=10

##MonteCarlo

#We create an empty vector to save the asset evolution
S=np.zeros(N+1)
```

```python
#We assign the initial value
S[0]=S0

#We run our MonteCarlo
for i in range (1,N+1):
        S[i]=S[i-1]*np.exp((r-0.5*sigma**2)*T/N+\
                        sigma*np.random.normal(0,1)*np.sqrt(T/N))

##Creating the replication portfolio

#We create an empty vector for the replication portfolio evolution
value=np.zeros(N+1)

#We create an empty vector for the option delta evolution
delta=np.zeros(N+1)

#We create an empty vector for the cash evolution
cash=np.zeros(N+1)

#We create an empty vector for the option value evolution
optionprice=np.zeros(N+1)

#The initial value of the option price is the BSM price
optionprice[0]=BS(S[0], K, T, r, sigma,"call" )

#The initial value of the replication portfolio is the option value
value[0]=optionprice[0]

#We calculate the delta
delta[0]=DeltaCall(S0,K,r,sigma,T)

#We calculate the cash
cash[0]=value[0]-delta[0]*S[0]

#For each time step
for i in range (1,N+1):

        #We calculate the replication portfolio
        value[i]=cash[i-1]*np.exp(r*T/N)+delta[i-1]*S[i]

        #We calculate the weights of the replication portfolio
        #if we are not at maturity time
        if i!=N:
            #The delta value
            delta[i]=DeltaCall(S[i], K, r, sigma, T*(N-i)/N)

            #We reassign the cash amount
            cash[i]=value[i]-delta[i]*S[i]
```

```
        #We calculate the option value.
        #Note that at maturity time, the value is the payoff
        #Note that the time to maturity decreases at each timestep
        if T*(N-i)/N==0:
            optionprice[i]=np.maximum(S[i]- K, 0)
        else:
            optionprice[i]=BS(S[i], K, T*(N-i)/N, r, sigma,"call" )

#We create a figure
plt.figure(figsize =(15 ,5))

#We set a title
plt.title("N="+str(N), fontsize=22)

#The x-axis title
plt.xlabel("Time", fontsize=18)

#The y-axis title
plt.ylabel("Value", fontsize=18)

#We do the plot of the replication portfolio evolution
plt.plot(value, "k--", label="replicating portfolio")

#We do the plot of the option evolution
plt.plot(optionprice,"k", label="option")

#We put the legend at the best place of the plot
plt.legend(loc="best", fontsize=14)

#We show the plot
plt.show()
```

9.3 CHAPTER'S DIGEST

The replication of an option by just assets and risk-free investment is a classical problem in finance. We have seen how to do this replication via Delta hedging, a technique consisting of reorganizing the replicating portfolio to have, at every moment, a quantity of assets equal to the Delta of the option. This technique can be applied to the binomial model and also to the Black-Scholes-Merton model. Notice that, in the Black-Scholes-Merton model, this replication cannot be exact, and this leads to a hedging error (that is, to a difference between the value of the option and the value of the replicating portfolio).

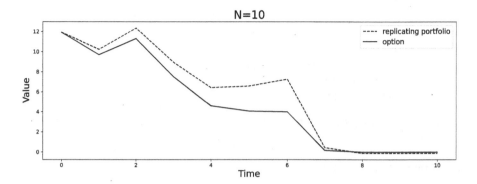

Figure 9.3 Simulation of the evolution of a call option price and the value of the corresponding replicating portfolio, with $N = 10$.

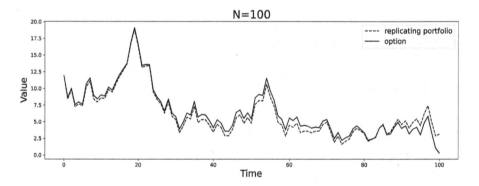

Figure 9.4 Simulation of the evolution of a call option price and the value of the corresponding replicating portfolio, with $N = 100$.

9.4 EXERCISES

- Take $r = 0.05$, $S_0 = 100$ and consider a vanilla call with $K = 90$ and $T = 1$. Imagine we replicate this call taking $N = 5$.

  If $S_{0.1} = 100$, $S_{0.2} = 100$, $S_{0.3} = 100$, $S_{0.4} = 100$ and $S_{0.5} = 100$,

  what is the final value of our replicating portfolio? What is the final value of the option? **Solution: At maturity, the value of the** replicating portfolio is **16.8, and the value of the option is 10.**

- Take $r = 0.05$, $S_0 = 100$ and consider a vanilla call with $K = 90$ and $T = 0.4$. Imagine we replicate this call taking $N = 4$.

  If $S_{0.1} = 90$, $S_{0.2} = 100$, $S_{0.3} = 90$, $S_{0.4} = 100$,

  what is the final value of our replicating portfolio? What is the final value of the option? **Solution: At maturity, the value of the replicating portfolio is 8.5, and the value of the option is 10.**

APPENDIX A

Introduction to Python

Since the beginning of trading and the use of derivatives in the modern era, computers have played an important role. Although today it is easy to establish a relationship between finance and computers, for example, for tasks related to data analysis or in the search for patterns in the evolution of stocks, one of its main advantages is speed. On the one hand, electronic commerce represented a great advance in the market, while on the other hand, it allows us to perform very complex calculations in a short time. For financial derivatives, this last advantage is very useful, especially for complex derivatives calculations.

This chapter is a basic introduction to Python, a programming language with which we can indicate a multitude of tasks to the computer with just a few lines of code. The use of this language is not random, or is it a whim of the authors. In the financial system, we can find other languages such as C# or C++ that may have other advantages, but Python is also used. It is a high-level language, which means that its use is closer to the user. In addition, every day is used more frequently for tasks related to data analysis, statistics, or Machine Learning. Therefore, it is a great tool for any student or person who wants to start programming.

A.1 BASIC OPERATIONS

In Python, we can give logical instructions to the computer. To start with, we'll use it as a calculator doing basic algebra.

Sum two numbers:

```
2+2
```

```
4
```

Multiply to numbers:

```
4*3
```

```
12
```

Divide two numbers:

```
4/3
```

```
1.3333333333333333
```

The integer part of a division:

```
4//3
```

```
1
```

The remainder of a division:

```
8%3
```

```
2
```

Calculate the power of a number:

```
#4^3
4**3
```

```
64
```

A.2 DATA TYPES

The data used in Python is classified internally by their type. For example:

- An integer has type int. For example, 7. It is the abbreviation for integer number.

- A decimal number has type float. For example, 8.2. It is the abbreviation for floating number.

- A character like 'c' or 'Barcelona' has type str. It is the abbreviation for string.

- A Boolean: True or False.

Python will assign the type automatically unless the user imposes a different kind.

A.3 VARIABLES

In any programming language, it is possible to store and use data by referring to it by a name chosen by the code developer. This object will be called a variable. We can define a variable as a container for storing data values. We can think of a variable as a name attached to a particular object stored in the memory of your computer.

In Python, variables do not need to be declared or defined beforehand, as is the case in other programming languages. To create a variable, we can assign the object to the name using a single equal sign (=). The operand to the left of the = operator is the variable's name the one to the right is the value stored in the variable. When we assign the value, we can use it in different code parts or an expression. The value can be changed during the program execution.

In a variable can be store any data type: a number, an array, a string, etc. A variable is created once you assign a value to it.

Example A.1

```
a=7
b=2
c=a+b
print(a+b)
print(c)
```

```
9
9
```

You can choose the name's variable you want but is recommended to follow the next rules:

- A variable name must be meaningful. The name should be descriptive and short.

- A variable name must start with a letter or the underscore character.

- A variable name can only contain alpha-numeric characters and underscores (A-z, 0-9 and)

- Variable names are case sensitive: stock and Stock are two different variables.

- Avoid reserved words, you can find them by typing help('keywords').

```
help("keywords")
```

Here is a list of the Python keywords. Enter any keyword to get more help.

False	class	from	or
None	continue	global	pass
True	def	if	raise
and	del	import	return
as	elif	in	try
assert	else	is	while
async	except	lambda	with
await	finally	nonlocal	yield
break	for	not	

A.4 PRINT

Sometimes, we may be interested in writing messages as an output. For example, to request a specific value in a variable or to debug parts of our code. This is done with the command 'print'. The most famous sentence in the world of programming is the following:

```
print("Hello, world!")
```

```
Hello, world!
```

If we define a variable with a name, we can personalize the message.

```
name="Marc"
print("Hello ", name , "!")
```

```
Hello, Marc!
```

From a technical point of view, we have encapsulated the name 'Marc' in a variable name of type string. Then, when we print the message, we recall the variable. As you can see, it is possible to alternate a text with some variables. Making the function very versatile.

```
age=42
print("I\'m ", age)
```

```
I'm 42
```

A.5 PACKAGES

Python is a 'package' based programming language. When we start writing our code, we can define variables and write code pieces, but it is possible to enrich the environment by using packages. Metaphorically, in Python, as in other languages, we have a library full of books with different specialties. A package is the equivalent of a book, where there are predefined functions that perform specific tasks: from creating a matrix, drawing a histogram, loading data ...

The most popular packages are:

- Numpy: The scientific computing library. It provides multidimensional arrays and matrices, along with a large collection of high-level mathematical functions to operate on these arrays.

- Pandas: It is a library for working with relational or labeled data. We will use to import data from different sources. It provides various data structures and operations for manipulating numerical data and time series.

- Matplotlib: It is a data visualization and graphical plotting library.

A.6 ROCKING LIKE A DATA SCIENTIST

In Finance, one of the most relevant skills is knowing how to work with data from various sources such as financial series and other types of information. In this section, we will learn how to import data. To do it, we will use the package 'pandas'. This package is specially designed for data manipulation and analysis.

A.6.1 Import Data

There are several ways to import data to Python. Some packages let us import data by code from different sources. Another option is to upload the data from Excel or a csv file.

We can import data in a csv format using the panda's function 'read_csv'. To use it, we need to import the package pandas, but also we need to tell the computer where is the function, so we need to write 'pandas.read_csv'. When we import the package we can give it a nickname, in this case it is usually 'pd', this will simplify the writing to 'pd.read_csv'.

```
#We import the package
import pandas as pd

#We load the data
data=pd.read_csv("filename.csv")
```

We are importing the data into the variable 'data'. The file should be in the same folder than the code. We can refer to any file in our file system by writing the full path.

```
import pandas as pd
data=pd.read_csv("C:/Path/filename.csv")
```

Note that when typing the path we use the delimiter '/', on some systems when copying the path the delimiter used is '\', so be careful when writing the path.

This function have several extra functionalities that will be helpful to import different types of files, a few of them:

- Specify the delimiter parameter between columns. The most common csv files separate the columns by a comma but it is not always the case. You can specify which is the delimiter of the file you can import.

  The common file will have the following format:

  ```
  Date,Close,Low,High
  2010-06-29,100,99.19,100.62
  2010-06-30,100.49,100.17,100.95
  ```

But sometimes the delimiter can be different, for example:

```
Date;Close;Low;High
2010-06-29;100;99.19;100.62
2010-06-30;100.49;100.17;100.95
```

In the second case, you will need to define the type of delimiter.

```
import pandas as pd
data=pd.read_csv("C:\Path\filename.csv", delimiter=";")
```

- Specify the decimal point or thousand point. Depending on where you are from, you will use a different decimal point. In Europe, ',' is used, such as '0,5', while elsewhere it is used '.', for example, '0.5'. If the decimal point is '.', the thousands point is ',' and vice versa. We can specify in what format the data comes.

```
import pandas as pd
data=pd.read_csv("C:\Path\filename.csv", delimiter=";",
            decimal=",", thousands=".")
```

- Specify if we want the headers or not. By default, the function will assume that the first row is the name's column. You can specify if there are no headers or what line they are on, but if they are on a different line than the start, none of the previous lines will be imported.

```
import pandas as pd
data=pd.read_csv("C:\Path\filename.csv", headers=None)
```

There are more functionalities such as choose the column's names, skip rows, define the type of each column, ...

There is a similar function to import data directly from Excel, 'read_excel', it works similarly.

A.6.2 Using Dataframes

In the previous section, we have learned how to import data and assign it to a variable. The data is imported into a Dataframe. But what is a Dataframe? A Dataframe is a two-dimensional data structure. In other words, it is a table with information.

Some of the benefits are:

- Having columns with different types.

- The size can change and columns can be added.
- Labelled axes. This can help us to have a more clear code.
- Can perform arithmetic operations on rows and columns.

> **Example A.2**
>
> We are going to import information of a stock.
>
> ```
> #Import Panda
> import pandas as pd
>
> #We import the data
> Stock=pd.read_csv("C:\Path\Stock.csv")
>
> #We print the data
> Stock
> ```

	Date	Close	Low	High
0	2010-06-29	100.00	99.19	100.62
1	2010-06-30	100.49	100.17	100.95
2	2010-07-01	99.22	97.82	101.02
3	2010-07-02	98.98	98.15	100.16
4	2010-07-06	101.44	100.54	101.56
...	...	...	...	...
3026	2022-07-07	533.22	531.68	535.28
3027	2022-07-08	526.51	524.93	527.03
3028	2022-07-11	533.33	528.72	539.91
3029	2022-07-12	536.45	533.69	540.00
3030	2022-07-13	529.51	525.29	532.18

3031 rows × 4 columns

Figure A.1 Dataframe.

It is possible to obtain a basic statistical description by using the property 'describe', see Figure A.2.

```
Stock.describe()
```

	Close	Low	High
count	3031.000000	3031.000000	3031.000000
mean	343.627562	341.050459	345.528987
std	187.591822	186.168873	188.622608
min	74.910000	74.460000	74.970000
25%	183.315000	181.870000	183.985000
50%	321.550000	320.130000	323.740000
75%	464.420000	460.035000	466.185000
max	944.330000	935.530000	953.480000

Figure A.2 Dataframe description.

We can refer to a column by their name.

```
Stock.Date
```

```
0        2010-06-29
1        2010-06-30
           ...
3029     2022-07-12
3030     2022-07-13
Name: Date, Length: 3031, dtype: object
```

Or to a specific point on the table.

```
Stock.Date[0]
```

```
'2010-06-29'
```

We can also do operations between columns.

```
(Stock.High+Stock.Low)/2
```

```
0          99.905
1         100.560
            ...
3029      536.845
3030      528.735
Length: 3031, dtype: float64
```

We can create a new column by referring to the dataframe.loc[:, "Name_new_column"], see Figure A.3. For example:

```
Stock.loc[:,"HighLow-Mid"]=(Stock.High+Stock.Low)/2
```

	Date	Close	Low	High	HighLow-Mid
0	2010-06-29	100.00	99.19	100.62	99.905
1	2010-06-30	100.49	100.17	100.95	100.560
2	2010-07-01	99.22	97.82	101.02	99.420
3	2010-07-02	98.98	98.15	100.16	99.155
4	2010-07-06	101.44	100.54	101.56	101.050
...	...	...	...	...	...
3026	2022-07-07	533.22	531.68	535.28	533.480
3027	2022-07-08	526.51	524.93	527.03	525.980
3028	2022-07-11	533.33	528.72	539.91	534.315
3029	2022-07-12	536.45	533.69	540.00	536.845
3030	2022-07-13	529.51	525.29	532.18	528.735

3031 rows × 5 columns

Figure A.3 Add a column.

If we want to calculate the return of a stock, we can use the 'shift' property that allows us to put the prices of the previous day, or any other past or future data, in the same row.

```
Stock.loc[:,"Return"]=Stock.Close/Stock.Close.shift(1)-1
```

A.6.3 Make Plot

A very important tool when working with data it is to visualize it. It helps us to understand and interpret it, but also to communicate it. It is easier for us to find relationships in the data when we view it graphically. For example, we can identify trends, patterns, frequency, outliers, ...

As we can imagine, there are special packages to do plots on Python, one of the most popular is 'matplotlib'. When are going to see different examples on how to use it.

```
#Import Package to do Plots
import matplotlib.pyplot as plt

#We write a Title for the Plot
plt.title("Prices of Stock")

#We write a x-axis label
plt.xlabel("Date")

#We write a y-axis label
plt.ylabel("Price")

#Do the Plot
plt.plot(pd.to_datetime(Stock.Date),Stock.Close)

#Show the Plot
plt.show()
```

Figure A.4 My first plot.

One of the first things we can be interested in is to modify the size of the plot. We need to add the command 'plt.figure(figsize =(x,y))' where x is the length and y the width, see Figure A.5.

```
#Import Package to do Plots
import matplotlib.pyplot as plt

#Modify the lenght and the width
plt.figure(figsize =(15,7))

#We write a Title for the Plot
plt.title("Prices of Stock",fontsize=16)

#We write a x-axis label
plt.xlabel("Date",fontsize=16)

#We write a y-axis label
plt.ylabel("Price",fontsize=16)

#Do the Plot
plt.plot(pd.to_datetime(Stock.Date),Stock.Close)

#Show the Plot
plt.show()
```

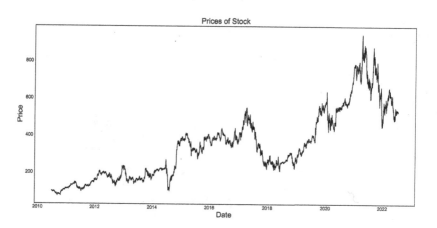

Figure A.5 Changing the size of a plot.

Sometimes, It is also interesting to draw two time series at the same time. In this case, we need to write the 'plot' instruction twice, see Figure A.6.

```
#Import Package to do Plots
import matplotlib.pyplot as plt

#Modify the lenght and the width
plt.figure(figsize =(15,7))

#We write a Title for the Plot
plt.title("Prices of Stock",fontsize=16)

#We write a x-axis label
plt.xlabel("Date",fontsize=16)

#We write a y-axis label
plt.ylabel("Price",fontsize=16)

#Plot the 1st time serie
plt.plot(pd.to_datetime(Stock.Date),Stock.Close,label="Close")

#Plot the 2nd time serie
plt.plot(pd.to_datetime(Stock.Date),0.5*Stock.Close,
        label="50% Close")

#Put the label on the best plot position
plt.legend(loc="best",fontsize=16)

#Show the Plot
plt.show()
```

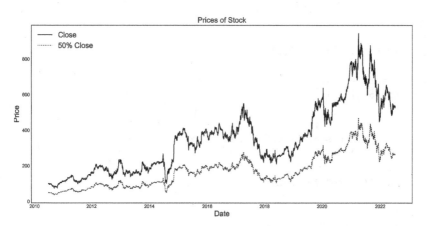

Figure A.6 Plot two time series.

Note that in this case, when doing the plot command we have added the label for each time series, so that it is possible to differentiate them.

204 ■ Introduction to Python

Sometimes it is helpful to compare different plots. To do that, the package 'matplotlib' let us to group plots in one figure. We will use the command plt.subplots(x,y,z) where x is the rows number, y is the column number and z the number of the plot, see Figure A.7.

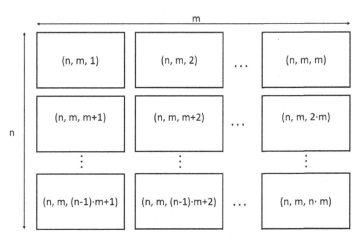

Figure A.7 How subplot works.

For example, if we plot the previous plot in two different graphics, see Figure A.8:

```
#Import Package to do Plots
import matplotlib.pyplot as plt

#Modify the length and the width
plt.figure(figsize =(15,7))

#We do a column with two plots. Here is the 1st.
plt.subplot(1,2,1)

#We write a Title for the Plot
plt.title("Prices of Stock",fontsize=16)

#We write a x-axis label
plt.xlabel("Date",fontsize=16)

#We write a y-axis label
plt.ylabel("Price",fontsize=16)

#Plot the 1st time serie
plt.plot(pd.to_datetime(Stock.Date),Stock.Close, ".", label="Close")

#Put the label on the best plot position
```

```
plt.legend(loc="best",fontsize=16)

#Here is the 2nd.
plt.subplot(1,2,2)

#Plot the 2nd time serie
plt.plot(pd.to_datetime(Stock.Date),0.8*Stock.Close,"-",
        label="80% Close")

#We write a Title for the Plot
plt.title("Prices of 80%Stock",fontsize=16)

#Put the label on the best plot position
plt.legend(loc="best",fontsize=16)

#Show the Plot
plt.show()
```

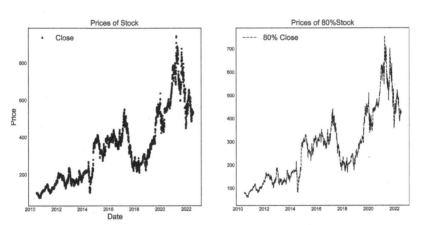

Figure A.8 Two plots in one figure.

The plot information such as the title, the x-axis, the y-axis, or the legend has to be defined for each subplot.

One the most typical graphics that we make are histograms. A histogram is a graphical representation that organizes a group of data points into ranges, also known as bins or buckets. In fact, it is an approximation of the empirical data distribution. To construct it, the first step is to determine the bucket range and then count how many values fall into each bucket, see Figure A.9.

```
#We import the package to do plots
import matplotlib.pyplot as plt
```

```
#We set the plot size
fig = plt.figure(figsize=(15,7))

#We define a cool plot style
plt.style.use("seaborn-white")

#We will do 50 bins
nbins=np.int(50);

#We do the histogram skipping the NA values.
plt.hist(Stock.Return.dropna(), bins=nbins,
        color = "gray", edgecolor = "black")

#We do the plot
plt.show()
```

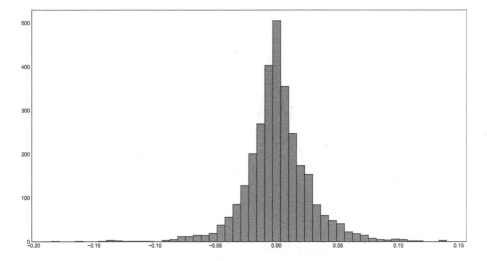

Figure A.9 A Histogram.

A.7 CHAPTER'S DIGEST

This chapter is a basic introduction to how to use Python. We have learned to do elementary mathematical operations. Also, we have focused on getting information from a csv file. Here, we have seen different options to upload the csv file. We have loaded the information into a variable called Dataframe. We have seen how we can refer to the data, a statistical description by column, or how to add new columns.

Furthermore, we have learned how to do plots. From the basic ones to a histogram. We have seen that we can make different plots in a figure and how it is organized in plot subsets.

A.8 EXERCISES

1. Download a stock's historical data of at least a period of one year. Then:

    (a) Get the description of the data.

    (b) Do a plot with the Stock close quotes.

2. Using the data of the previous exercise:

    (a) Add a column in the dataframe with the Log-Returns.

    (b) Calculate the daily mean and standard deviation.

    (c) Make a histogram of the Log-Returns.

    (d) Make a graphic with the following 2 plots:
    - Stock historical evolution.
    - Stock Log-Returns.

3. Download the historical data of the last year for a different stock. Do a plot with the evolution of both stocks with labels.

APPENDIX B

Introduction to Coding in Python

In the previous Appendix, we learned how to load data and use some default Python functions. Although there is an extensive collection of packages and functions, it is important to learn to define your own functions. This chapter will introduce functional programming concepts.

B.1 DEFINE YOUR OWN FUNCTIONS

We are going to learn how to create our own functions. This will allow us to create specific solutions for different problems, but also to use and reuse them more than once. The idea behind this is to create small functions, modules, that are called through the main code. This allows each part of the code to be reviewed at an 'atomic' level.

But what is a function? A function is a code block, also named script, that only executes when it is called. The function name has to be unique and, although it is not necessary, but if convenient, the name has to be descriptive of what it does.

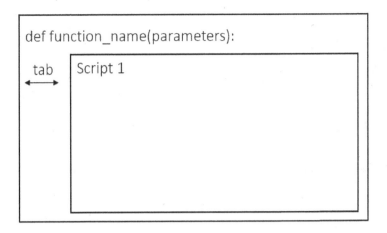

Figure B.1 Definition structure.

For creating a function, the first step is to write the function header. We will use the command 'def' to indicate that we are defining a function. Then, we will have to choose the function name followed by the parameters list in parentheses. Once we have written them all, we use the colon (:) to mark the end of the function header, see Figure B.1. After writing the function header, we will write a code block separated by a tabulation. Python uses tabbing to determine the level of dependency for statements. So all the instructions that are part of our function must be tabulated in at least one position. The computer will understand that everything behind the function header and tabbed will be part of the function.

Example B.1

Let's going to see a basic example.

```
def say_hello():
  #This function says Hello!
  print("Hello")
```

```
say_hello()
Hello!
```

This function will print 'Hello!' every time that runs.

The example above is basic. It doesn't seem too practical. We can generalize the process. Using one input, we can get a custom output. For example, adding the user name as a parameter.

Example B.2

Let's going to complicate the example B.1.

```
def say_hello(name):

    #This function says Hello using the value of the local variable
    #"name"

    print("Hello" , name , "!")
```

```
say_hello("Adele")
Hello Adele!
```

This function will print Hello with a name depending on the input variable 'name'. The variable 'name' is a local variable, only exists inside the function 'say_hello'.

We can make the code more complex. We can add more than one variable.

Example B.3

We are going to create a script based on the example B.1, but this time we will add a second variable referring to the surname.

```
def say_hello(name, surname):

    #This function says Hello using the value on the local variables:
    #"name" and "surname"

    print("Hello" , name, surname, "!")
```

```
say_hello("Adele", "Smith")
Hello Adele Smith!
```

B.2 IF

It seems that the code executes sequentially all the tabbed lines after the definition. This is not always true. Conditionals can be added to run only chunks of code. One of the most popular is the 'If' statement. This statement is used for decision making. If the condition of the declaration is true, the code tabulated below is executed (see Figure B.2).

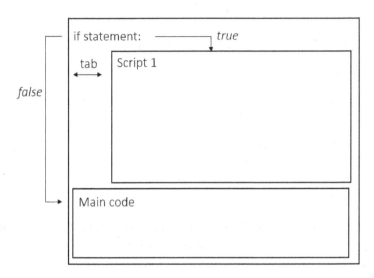

Figure B.2 Basic If Statement.

Let's going to see a basic example.

Example B.4

A basic if statement example.

```
def is_positive(number):

  #This function tells you if a number is positive

  if number > 0:
    print("The ", number, " is a positive number.")
  print("This is always printed.")
```

Using the function with the number 5, we obtain the following result.

```
is_positive(5)
The  5 is a positive number.
This is always printed.
```

But, using the function with the number −1, we obtain the following result.

```
is_positive(-1)
This is always printed.
```

We have seen the simplest expression of the 'If' statement. We can add the 'else' statement. In this case, if the condition is true, a script is executed. If it is false, another script is executed and then continues with the main code (see Figure B.3).

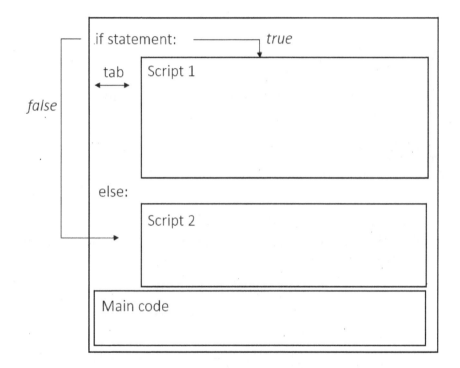

Figure B.3 If-Else Statement.

Example B.5

We are going to improve the code of the example to include an else condition.

```
def sign(number):

    #This function tells you if a number is positive or negative
```

```
if number >= 0:
  print("The ", number, " is a positive number or zero.")
else
  print("The ", number, " is a negative number.")
print("This is always printed.")
```

If we use the function with the number 5, we obtain the following result.

```
sign(5)
The  5 is a positive number.
This is always printed.
```

But, if we use the function with the number −1, we obatin the following result.

```
sign(-1)
The -1 is a negative number.
This is always printed.
```

We have included zero with all the positive numbers. We might want to differentiate this case. We can add more conditions with an 'elif' statement (see Figure B.4).

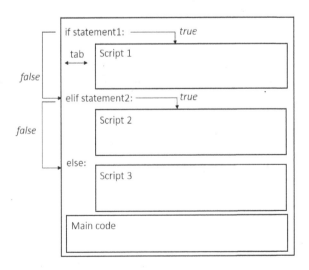

Figure B.4 If-Elif-Else Statement.

Example B.6

We are going to improve the code of the example to include an elif condition.

```
def sign(number):

  #This function tells you if a number is positive,
  #negative or zero

  if number > 0:
    print("The ", number, " is a positive number or zero.")
  elif number==0:
    print("This is zero!")
  else:
    print("The ", number, " is a negative number.")
  print("This is always printed.")
```

If we use the function with the number 5, we obtain the following result.

```
sign(5)
The  5 is a positive number.
This is always printed.
```

If we use the function with the number -1, we obtain the following result.

```
sign(-1)
The -1 is a negative number.
This is always printed.
```

But, if we pass 0 to the function, we have the result.

```
sign(0)
This is zero!
This is always printed.
```

B.3 FOR

We have seen that we can teach computers to perform a task. Computers are very efficient at quickly performing repetitive tasks. One of the first repetitive tasks we learn to do as children is to count. Many processes depend on counting. The Python statement for count is for (see Figure B.5).

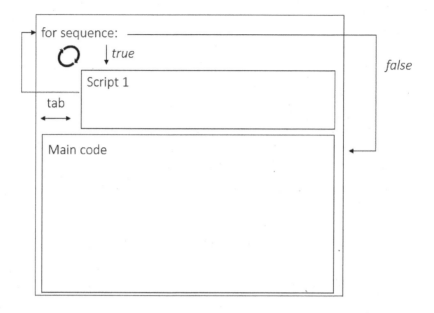

Figure B.5 For Statement.

Example B.7

We are going to see a basic loop example.

```
def counting(start, stop):

  #This function counts from start to stop (not included)

  for i in range(start, stop):
    print(i)
  print("This is always printed.")
```

We execute the counting function using start=1 and stop=4.

```
counting(1, 5)
1
2
3
4
```

We are telling the computer to repeat the print function for each i of the iteration. The statement stops before reaching 5. The print function depends on the for statement, that is because it is tabulated.

Example B.8

The factorial of a positive integer n, denoted by n!, is the product of all positive integers less than or equal to n.

$$n! = n \cdot (n-1) \cdot (n-2) \cdot \ldots 2 \cdot 1$$

```
def factorial(n):

  #This function calculates the factorial

  #The initialise the variable factorial
  factorial=1;

  #By using the for iteratively, we compute the factorial
  for i in range(2,n+1):
    factorial=factorial*i

  print(factorial)
```

We execute the factorial function with n=3.

```
factorial(3)
6
```

We execute the factorial function with n=10.

```
factorial(10)
3.628.800
```

B.4 CREATING MATRICES

When making code, sometimes we have a lot of information that we need to allocate. We have learned how to use variables and Dataframes, but there is another possibility, create matrices. Matrices are very important data structures for many mathematical and scientific calculations. In Python, there is not a Matrix type, we can create a matrix as a vector of vectors, or in better words, a list of lists.

Example B.9

```python
#We create a Matrix as a list of list, i.e. vector by vector
A = [[1, 2, 3],
     [-4, -5, -6],
     [7, 8, 9]]

#We can print the whole structure
print("A =", A)

#Chose the 2nd row
print("A[1] =", A[1])

#Chose an element, for example, 3rd element 3rd row
print("A[2][2] =", A[2][2])

#Chose the last element of the 2nd row
print("A[0][-1] =", A[1][-1])

#We create a new empty list
column = [];
for row in A:

    #We allocate the second column row by row
    column.append(row[2])

#We print the 3rd column
print("3rd column =", column)
```

```
A = [[1, 2, 3], [-4, -5, -6], [7, 8, 9]]
A[1] = [-4, -5, -6]
A[2][2] = 9
A[0][-1] = -6
3rd column = [3, -6, 9]
```

We can access any value in the matrix. Positions start at position 0. The first position refers to the rows and the second to the columns. If we move from left to right or top to bottom, the number of positions increases. It also moves from right to left or bottom to top, but in this case, we use negative numbers starting by -1.

Here, we can see the positions of a matrix 3x3:

$$\begin{pmatrix} [0,0] & [0,1] & [0,2] \\ [1,0] & [1,1] & [1,2] \\ [2,0] & [2,1] & [2,2] \end{pmatrix}$$

and the positions counting from right to left and bottom to top:

$$\begin{pmatrix} [-3,-3] & [-3,-2] & [-3,-1] \\ [-2,-3] & [-2,-2] & [-2,-1] \\ [-1,-3] & [-1,-2] & [-1,-1] \end{pmatrix}$$

In the Numpy package, there is a special type of object that is more efficient for calculations. It is the array types.

Example B.10

Creating my first array.

```
#Load numpy package
import numpy as np

#Create an array
A = np.array([[1, 2, 3],
    [-4, -5, -6],
    [7, 8, 9]])

#Print the matrix
print("A=",A)

#Print the second column
print("2nd column", A[:,1])
```

```
A= [[ 1  2  3]
 [-4 -5 -6]
 [ 7  8  9]]
2nd column [ 2 -5  8]
```

Creating an array type Matrix is as difficult as create a Matrix as a list of lists. By changing the type to array, we can directly extract the second column without creating a new list and append the values for each row.

There is another option to create a matrix using Numpy. It is create a matrix full of zeros.

Example B.11

Creating a matrix full of zeros.

```
#Load numpy package
import numpy as np

#Create matrix full of zeros
A = np.zeros((3,3))

#Print the matrix
print("A=",A)
```

```
A= [[0. 0. 0.]
 [0. 0. 0.]
 [0. 0. 0.]]
```

Alternatively, we can create a matrix filled with ones or with only values on the diagonal.

Example B.12

```
#Load numpy package
import numpy as np

#Create matrix full of ones
A = np.ones((3,3))

#Print the matrix
print("A=",A)

#Create matrix with ones on the diagonal
B = np.diag(np.ones(3))

#Print the matrix
print("B=",B)
```

```
A= [[1. 1. 1.]
 [1. 1. 1.]
 [1. 1. 1.]]
B= [[1. 0. 0.]
 [0. 1. 0.]
 [0. 0. 1.]]
```

Sometimes we need to know dimensions of the matrix. If we have created the matrix using numpy, we can use the shape property by just using the command name_matrix.shape. Another import thing, is to learn to move within the matrix. We have learned that the for statement was important for counting, now we are going to see a practical example about how to use it with matrices.

Example B.13

```
#Load numpy package
import numpy as np

#Create matrix full of ones
A = np.zeros((3,4))

#Print the matrix
print("A=",A)

#Print the shape of the matrix
print("The matrix shape is", A.shape)

#We define the variable count
count=1

#For each row
for i in range(0, A.shape[0]):

    #For each position
    for j in range(0, A.shape[1]):

        #Write in the position [i,j] the variable count
        A[i,j]=count

        #Sum 1 to count
        count=count+1

#Print the matrix
print("A=",A)
```

```
A= [[0. 0. 0. 0.]
 [0. 0. 0. 0.]
 [0. 0. 0. 0.]]
The matrix shape is (3, 4)
A= [[ 1.  2.  3.  4.]
 [ 5.  6.  7.  8.]
 [ 9. 10. 11. 12.]]
```

What the double for statement is doing?

```
count=1
i is between [0,3)
j is between [0,4)

The loop for the variable i starts. The loop for the variable j
    starts.

i=0   j=0   A[0,0]=1    count=count+1=2
i=0   j=1   A[0,1]=2    count=count+1=3
i=0   j=2   A[0,2]=3    count=count+1=4
i=0   j=3   A[0,3]=4    count=count+1=5

The variable j is outside its range. The loop is over.
The variable i increase one position.
The loop for j starts again.

i=1   j=0   A[1,0]=5    count=count+1=6
i=1   j=1   A[1,1]=6    count=count+1=7
i=1   j=2   A[1,2]=7    count=count+1=8
i=1   j=3   A[1,3]=8    count=count+1=9

The variable j is outside its range. The loop is over.
The variable i increase one position.
The loop for j starts again.

i=2   j=0   A[2,0]=9    count=count+1=10
i=2   j=1   A[2,1]=10   count=count+1=11
i=2   j=2   A[2,2]=11   count=count+1=12
i=2   j=3   A[2,3]=12   count=count+1=13
```

B.5 CHAPTER'S DIGEST

This chapter is a basic introduction to coding. We have learned to define our functions. This will help us create our processes that will do exactly what we want. It is a way to automate actions that we will need to repeat and repeat. For example, creating our pricing functions. To enrich our functions, we have seen some examples of clauses like if...else and for, although there are many more. The first one is very useful because we can code in blocks and activate that part of the code based on a variable. The second will help us to repeat an action or move through different scenarios. Also, we have introduced how to create a matrix or a vector. In our case, it will be crucial to create our own Binomial model or to store the different stochastic paths of an asset in a MonteCarlo scheme.

B.6 EXERCISES

1. Define a function that given two number, inform you which one is the biggest. (Hint: use the clause 'if')

2. We can determine if a number is even or odd looking at the rest of a division. The rest of a division in Python is obtained with '%'. For example, the rest of 14/5 is obtained by '14%5'. Do a function that determine if an input is even or odd.

3. A water company is implementing a new bill system. For each house the following invoice is made

    - The first 30 liters of water are free.
    - From 30 liters to 150 liters, the cost per liter is 0.15 euros.
    - From 150 liters, the cost per liter is 0.25 cents.
    - The minimum invoice is of 6 euros.

    Can you define a function to calculate the cost for a given amount of liters spent. (Hint: use the clause 'if')

4. Create a vector of 5 position and numerate each one of the positions. (Hint: use the clause 'for')

5. Create a matrix of size 5 × 5. At each diagonal position, store the value 1. This is call Identity Matrix. (Hint: use the clause 'for')

6. Create a matrix of size 5 × 5. If you are in the diagonal position, store 1, otherwise store 2. (Hint: use the clause 'for' inside clause 'for')

Bibliography

ALGIERI, B (2018) A Journey Through the History of Commodity Derivatives Markets and the Political Economy of (De)Regulation. DOI 10.22004/ag.econ.281139

ALÒS, E AND GARCÍA LORITE, D (2021) Malliavin Calculus in Finance: Theory and Practice. Chapman and Hall/CRC

BACHELIER, L (1900) Théorie de la spéculation. PhD thesis, Paris: Gauthier-Villars

BERGOMI, L (2016) Stochastic Volatility Modeling. Chapman and Hall/CRC Financial Mathematics Series, Chapman and Hall/CRC, Boca Raton

BLACK, F (1976) The pricing of commodity contracts. Journal of Financial Economics. DOI https://doi.org/10.1016/0304-405X(76)90024-6

BLACK, F S AND SCHOLES, M S (1973) The pricing of options and corporate liabilities. Journal of Political Economy. DOI 10.1086/260062

BRENT, RP (1973) Algorithms for minimization without derivatives. Prentice-Hall

COX, JC AND ROSS, S A AND RUBINSTEIN, M (1979) Option pricing: A simplified approach. Journal of Financial Economics. DOI https://doi.org/10.1016/0304-405X(79)90015-1

DUPIRE, B (1994) Pricing with a smile. Risk

GARMAN, M B AND KOHLHAGEN, S W (1983) Foreign currency option values. Journal of International Money and Finance. DOI https://doi.org/10.1016/S0261-5606(83)80001-1

HEBER, G AND LUNDE, A AND SHEPARD, N AND SHEPPARD, K (2009) Oxford-man institute's realized library, version 0.3. Oxford-Man Institute, University of Oxford.

KASSOUI, S T AND THORP, E O (1967) Beat the Market. Random House

LONGSTAFF, F AND SCHWARTZ, E (2001) Valuing american options by simulation: A simple least-squares approach. Review of Financial Studies. DOI 10.1093/rfs/14.1.113

MERTON, R C (1973) Theory of rational option pricing. The Bell Journal of Economics and Management Science. DOI 10.2307/3003143

OSBORNE, MFM (1959) Brownian motion in the stock market. Operations Research

SHELDON M ROSS (2014) Introduction to Probability Models, eleventh edn. Academic Press

SAMUELSON, P (1965) Rational theory of warrant pricing. Industrial Management Review

SPRENKEL, C (1961) Warrant prices as indications of expectations. Yale Economic Essays

Index

at-the-money, 27
    forward, 27
    spot, 27

bear spread, 36
Binomial model, 177
    Multi-Period, 69, 126
    Single-Period, 55
Black-Scholes
    equation, 114
Black-Scholes-Merton
    dividends, 140
    drawbacks, 141
    Exchange rates, 140
    formula, 116, 120, 124, 126
    framework, 113
    Futures, 141
    model, 109, 177, 181
Brownian Motion, 120, 159
Brownian motion, 159
bull spread, 36

forward
    contract, 9
    long contract price, 21
    long contract price (known income), 22
    long contract price (proportional yield), 22
    price, 11, 15
    price (known income), 18
    price (proportional yield), 20
    short contract price, 21
futures
    contract, 11
    price, 11

Greeks, 87, 127
    delta, 59, 60, 90, 128, 178, 181
    gamma, 90, 134
    rho, 92
    theta, 91, 132
    vega, 91, 137

hedging
    dynamic, 179
    error, 181
    static, 179

in-the-money, 27
interest rates, 30
intrinsic value, 27

law of Large Numbers, 151

margin
    account, 12
    call, 12
    initial, 12
    maintenance, 12
Monte Carlo method, 148, 155

options
    American, 25, 84, 158
    Asian, 51
    Asian (fixed), 51
    Asian (floating), 51
    asset-or-nothing, 45
    barrier, 48, 160
    barrier (knock-in), 48

barrier (knock-out), 48
barrier knock-out, 158, 161
basket, 52
bermudan, 53
binary, 45
Call, 25, 116
cash-or-nothing, 45
cliquet, 46
compound, 47
European, 25, 177, 181
Exotics, 25
forward start, 46
lookback, 50
lookback (fixed), 50
lookback (floating), 50
moneyness, 116
path-dependent, 47
Put, 25
put, 117
spread, 52
spread option, 162
spread options, 163
Vanilla, 25
out-the-money, 27

payoff, 10
position
    long, 9, 25
    short, 9, 25
put-call parity, 33, 172
Python
    Dataframe, 197
    definition, 209
    for clause, 215
    histogram, 205
    if clause, 211
    matrices, 217
    packages, 195
    plot, 201
    print, 194

variable, 193

replication portfolio, 60, 114, 177
risk-neutral, 64, 113

strike, 25, 29

Taylor series, 88, 89, 115
time to maturity, 31
time value, 27

volatility, 31
    historical, 170
    implied, 172, 174
    skew, 173
    smile, 173
    spot, 171, 174

Wiener process, 120